Lecture Notes in Electrical Engineering

Volume 316

More information about this series at http://www.springer.com/series/7818

About this Series

"Lecture Notes in Electrical Engineering (LNEE)" is a book series which reports the latest research and developments in Electrical Engineering, namely:

- Communication, Networks, and Information Theory
- Computer Engineering
- Signal, Image, Speech and Information Processing
- Circuits and Systems
- Bioengineering

LNEE publishes authored monographs and contributed volumes which present cutting edge research information as well as new perspectives on classical fields, while maintaining Springer's high standards of academic excellence. Also considered for publication are lecture materials, proceedings, and other related materials of exceptionally high quality and interest. The subject matter should be original and timely, reporting the latest research and developments in all areas of electrical engineering.

The audience for the books in LNEE consists of advanced level students, researchers, and industry professionals working at the forefront of their fields. Much like Springer's other Lecture Notes series, LNEE will be distributed through Springer's print and electronic publishing channels.

Mohsen A.M. El-Bendary

Developing Security Tools of WSN and WBAN Networks Applications

Springer

Mohsen A.M. El-Bendary
Helwan University
Cairo, Egypt

ISSN 1876-1100 ISSN 1876-1119 (electronic)
ISBN 978-4-431-56357-0 ISBN 978-4-431-55069-3 (eBook)
DOI 10.1007/978-4-431-55069-3
Springer Tokyo Heidelberg New York Dordrecht London

Printed on acid-free paper

Springer is part of Springer Science+Business Media (www.springer.com)

To my mother and father soul & family

I dedicate this book to my parents,
my wife and my children who supported
me all the time with their love and pray.

Mohsen A.M. El-Bendary

Preface

This book introduces wireless sensor networks (WSNs), the appearance of which has been made possible by the tremendous development of wireless technology (WT). The book presents the two most familiar types of networks, which are the WSNs and wireless body area networks (WBANs). These networks are configured and established for specific purposes:- sensing and monitoring the volume of something and transferring the captured signal through an air interface. The book thus presents these networks under the category of "object oriented wireless networks". Certain factors distinguish these networks from other wireless networks. A sensing network is a wireless network and the nodes that merged with the interface connect it to the surrounding medium.

The book also focuses on environmental application because the WSN and the environment are strongly related. Environmental monitoring is one of the most important applications of the WSN. It may include sensing the earth's surface activities such as earthquakes and volcanic activity, and it may also, be used for the detecting forest fire and monitoring plants and animals.

One capability of the WSN that distinguishes it from the human eye is that it can sense unseen events. It is provided with an arm to touch surrounding objects, making it an interesting search device. The book presents the WSN from this angle, i.e., that it may be enable the scientists to discover as-yet unknown natural phenomenon. Although, most of researchers treat the two familiar WT as different networks, there are some shared features between the WBANs and WSNs. The first is locally applicable wireless medical applications. The second goes beyond local applications, with field applications that extend to global scale.

The security of the WSNs is a critical issue. WSN application scenarios may be related to national security in specific fields such as military applications and detection of nuclear radiation for example. The security of WSN is the most important concern for the basic functionality of networks. The availability of network services, confidentiality and integrity of the data can be achieved by assuring that security issues have been properly dealt with. In general, the WSNs often suffer from attacks because of its features like open mediums, dynamic changes in topology, lack of central monitoring and management, cooperative algorithms and no clear defense mechanisms. These factors have changed the battle field for the WSN against security threats. Therefore, this book introduces proposed cooperative algorithms between the WSN nodes as well as features of sensing station mobility.

Scenarios are presented for developing WSN security using watermarking techniques and other, different scenarios are also presented.

The aim of this book is to convey to readers the knowledge that is essential for a complete understanding of WSNs with a focus on the wireless personal area networks (WPANs) that are widely utilized for implementing the WSN applications scenario. For example, Bluetooth technology is considered the engine of personal communications development; in general WPAN technology can be regarded as the engine for spreading the development of WT applications. In addition, it opens the door to the question of it is used developmentally.

This is the first book about wireless communication that gives a comprehensive introduction to data-length complexity and the computational complexity of processed data and error control schemes, which are essential points in system design especially in pre-determined wireless network applications. With regarding to complexity, the following are covered; definition of the computational complexity of communications systems and wireless networks; and the degree of complexity which, because of the amount of processed data transmitted over the networks, is mostly ignored in the evaluation complexity. The length of transmitted data is summarized in the data-packet length and the length of other components of packets such as the header and tail. The type of error control coding schemes contributes to increasing the complexity according to the complexity of the encoder and decoder of this error control scheme.

This book will be helpful to all academic and technical staff in the fields of telecommunications and wireless communications, where it presents many scenarios for enhancing weak error-control performance and other scenarios for complexity reduction of the wireless data and image transmission through the techniques proposed. Many examples are given to help the readers to understand the material covered in the book. Additional resources such as the MATLAB codes for some of the examples are included in the book.

Cairo, Egypt Mohsen A.M. El-Bendary

Contents

About the Author

Name: MOHSEN AHMED MAHMOUD MOHAMED KASSEM EL-BENDARY.
Date of birth: 20/9/1975.
Phone: 002-02-33266239
Mobile: 0020-128-8712859
0020-106-0747274
E-Mail: **dr_m.kassem@yahoo.com**
Engmohsen2004@yahoo.com

Affiliation: Department of Electronics and Communications Technology, Faculty of Industrial Education, Helwan University, Cairo, Egypt.

Qualifications:

1. BSc. in Electrical Communications, Faculty of Electronic Engineering, Menoufia University, May 1998.
2. MSc. in Electrical Communications "**Improving Performance Bluetooth Networks through Channel Codding**", Faculty of Electronic Engineering, Menoufia University, 2008.
3. PhD. in Electrical Communications "**Efficient Multimedia Transmission over Bluetooth System**", Faculty of Electronic Engineering, Menoufia University, 2012.

Areas of Interests:

1. Wireless networks, specially. The wireless personal communications such as WPANs, WSN, and WBANs.
2. Image Processing: Enhancement of restoration of degraded and noisy images, multi channel image processing, error of image concealment, color image processing, image watermarking, encryption, and data hiding).
3. Computational complexity of Channel coding.
4. Implementation the different security systems such as Fire Alarm and Access control also HVAC system using the wireless technologies.

Other Practical Activities:

- 12 years experience Design details, installation and installation supervision, testing and commissioning, and programming for the light current systems.
- Participation in many projects, installation and installation supervision, during the last 12 years, Hotels, Universities, Airports, Hospitals, and Petroleum Companies sites.

Other Scientific Activities:

- He served as reviewer for scientific papers in the fields of signal and image processing for the number of International Journals in communications and electronics such as the International Journal of Electronics and Journal of IEEE Transactions on Vehicular Technology.

Books Publications:

- Book Title: "**Art of Error Recovery Techniques over Wireless Networks**", LAP Lambert Academic Publishing, ISBN-978-3-659-28008-5, 10 Oct. 2012.
- Book Title: "**WPANs Technologies Modeling and Simulation**", LAP Lambert Academic Publishing, ISBN-978-3-659-25440-6, 2013.
- Book Title: "**Definition of Nano World through Question and Answer**" in Arabic Language, Under Publishing Process.

Journal Publications:

1. **M. A. M. Mohamed**, A. Abou El-Azm, N. A. El-Fishawy, M. A. R. El-Tokhy and F. E. Abd El-Samie "Optimization of Bluetooth Frame Format for Efficient Performance," **PIER**-B, Progress In Electromagnetics Research M, Vol. 1, 101–110, 2008.
2. **M. A. M. Mohamed**, A. E. Abou El-Azm, N. A. El-Fishawy, M. A. R. - El-Tokhy, F. Shawky and F. E. Abd El-Samie, "Bluetooth Performance Improvement Using Convolutional Codes," Journal of Automation & Systems Engineering, Vol. 3, No. 1, 2009.
3. W. Al-Nuaimy, **M. A.M. El-Bendary**, A. Shafik, F. Shawki, A.E. Abou-El-azm, N.A. El-Fishawy, S. M. Elhalafawy, S. M. Diab, B. M. Sallam, F. E. Abd

El-Samie, and H.B. Kazemian, "An SVD Audio Watermarking Approach Using Chaotic Encrypted Images," **DSP**, Digital Signal Processing, Vol. 21, No. 6, pp. 764–779, 2011, **Elsevier** DOI:10.1016/jdsp.2011.01.013 (**Impact Factor: 1.617**).

4. **M. A. M. M. El-Bendary**, A. Abou El-Azm, N. El-Fishawy, F. S. M. -Al-Hosarey, M. A. R. Eltokhy, F. E. Abd El-Samie, and H. B. Kazemian, "An Efficient Chaotic Interleaver for Image Transmission over IEEE 802.15.4 Zigbee Network" Journal of Telecommunications and Information Technology **JTIT**, Vol. 3, No.2, pp. 67–73, 2011.
5. **M. A. M. El-Bendary**, A. E. Abou-El-azm, N. A. El-Fishawy, F. Shawki, M. El-Tokhy, F. E. Abd El-Samie, H. B. Kazemian, "Enhancing the Image Transmission over Wireless Networks through a Novel Interleaver" **KSII Transactions on Internet and Information Systems** Vol. 5, No. 9, pp. 1528–1543, 2011, DOI: 10.3837/tiis.2011.09.003, (**Impact Factor: 0.20**).
6. **M. A. M. El-Bendary**, A. Abou El-Azm, N. El-Fishawy, F. S. M. Al-Hosarey, M. A. R. Eltokhy, F. E. Abd El-Samie, and H. B. Kazemian, "SVD Audio Watermarking: A Tool to Enhance the Security of Image Transmission over ZigBee Networks", Journal of Telecommunication and Information Technology, **JTIT**, pp. 99–107, Vol. 4, 2011.
7. **M. A. M. El-Bendary**, A. E. Abou-El-azm, N. A. El-Fishawy, F. Shawki, F. E. Abd El-Samie, H. B. Kazemian, "Embedded Throughput Improving of Low-rate EDR Packets for Lower-latency", **IJICE** International Journal of Information and Communications Engineering, pp. 238–247, Vol. 6:4, 2010, ISSN: 2010–4049, e-ISSN: E-ISSN: 201004049.
8. **M. A. M. El-Bendary**, A. E. Abou-El-azm, N. A. El-Fishawy, F. Shawki,, F. E. Abd El-Samie, H. B. Kazemian, "Image Transmission in Low-Power Networks in Mobile Communications Channel", **WASC**, World Academy of Science, Engineering and Technology, pp. 1019–1025, Vol. 80 2011, ISSN: 2010-376x, e-ISSN: 2010–3778.
9. **M. A. M. El-Bendary**, A. Abou El-Azm, N. El-Fishawy, F. S. M. Al-Hosarey, M. A. R. Eltokhy, F. E. Abd El-Samie, and H. B. Kazemian, "Performance of the Audio Signals Transmission over Wireless Networks with the Channel Interleaving Considerations," **EURASIP-ASMP, Journal on Audio, Speech, and Music Processing**, **Springer** Vol. 4, 2012. DOI:10.1186/1687-4722-2012-4, (**Impact Factor = 0.71**).
10. **M. A. M. El-Bendary**, A. Abou El-Azm, N. El-Fishawy, F. S. M. Al-Hosarey, M. A. R. Eltokhy, F. E. Abd El-Samie, and H. B. Kazemian, "Image Transmission over Mobile Bluetooth Networks with Enhanced Data Rate Packets and Chaotic Interleaving," **Wireless Networks**, **WINET**, **Springer**, DOI 10.1007/s11265-010-0516-3, (**Impact Factor =0.52**).
11. **M. A. M. El-Bendary,** A. Abou El-Azm, N. El-Fishawy, F. S. M. Al-Hosarey, M. A. R. Eltokhy, F. E. Abd El-Samie, and H. B. Kazemian, "JPEG image transmission over mobile network with an efficient channel coding and interleaving", **IJE, International Journal of Electronics** (**Impact Factor =2.923**).

12. **M. A. M. El-Bendary**, A. Abou El-Azm, N. El-Fishawy, F. S. M. Al-Hosarey, M. A. R. Eltokhy, F. E. Abd El-Samie, and H. B. Kazemian, "An Efficient Data Randomization Tool For Image Transmission over Mobile WPANs," Accepted for publication in the **International Journal of Communication System** (**IJCS**) (**Impact Factor =0.24**).

Conference Publications:

1. M. A. Mohamed El-Bendary, A. E. Abou-El-azm, N. A. El-Fishawy, A. Hagagg, F. Shawki and F. E. Abd El-Samie, "Real-Time Audio Signal Transmission Over The ACL Link In Bluetooth Systems" Proceedings of the URSI National Radio Science Conference (NRSC), Cairo, Egypt, March 2010.
2. M. A. Mohamed El-Bendary, A. E. Abou-El-azm, N. A. El-Fishawy, M. A. R. El-Tokhy, F. Shawki and F. E. Abd El-Samie, "Throughput Improvement over Bluetooth System Through Adaptive Packets", The 2009 International Conference on Computer Engineering & Systems (ICCES'09), December 14–16, 2009, Triumph Hotel, Cairo, Egypt.
3. M. A. Mohamed El-Bendary, A. E. Abou-El-azm, N. A. El-Fishawy, M. A. R. El-Tokhy, F. Shawki and F. E. Abd El-Samie, "Improving Power Efficiency of Bluetooth Systems with EDR Packets and Efficient Channel Coding", The 2009 International Conference on Computer Engineering & Systems (ICCES'09), December 14–16, 2009, Triumph Hotel, Cairo, Egypt.
4. M. A. Mohamed El-Bendary, A. E. Abou-El-azm, N. A. El-Fishawy, M. A. R. El-Tokhy, F. Shawki and F. E. Abd El-Samie,"Novel Technique for Improving Bluetooth Networks Security Through SVD-Based Audio Watermarking", The 2011 I.E. International Union of Radio Science, URSI GASS, Istanbul, Turkey, Aug. 12–20, 2011.
5. M. A. Mohamed El-Bendary, A. E. Abou-El-azm, N. A. El-Fishawy, M. A. R. El-Tokhy, F. Shawki and F. E. Abd El-Samie,"Power-Efficient of Image Transmission over Bluetooth Systems Using Randomization Technique", E-Medisys 2010 3rd International Conference: E-medical Systems, May 12–14, 2010 –Morocco.
6. M. A. Mohamed El-Bendary, A. E. Abou-El-azm, N. A. El-Fishawy, M. A. R. El-Tokhy, F. Shawki and F. E. Abd El-Samie, "Studying the Performance of Image Transmission over Mobile Bluetooth Network' through Using EDR Packets", E-Medisys 2010 3rd International Conference: E-medical Systems, May 12–14, 2010 –Morocco.
7. M. A. Mohamed El-Bendary, A. E. Abou-El-azm, N. A. El-Fishawy, M. A. R. El-Tokhy, F. Shawki and F. E. Abd El-Samie, "Performance of Image Transmission over IEEE 802.15.1 and IEEE 802.15.4 with Fragmentation Consideration", E-Medisys 2010 3rd International Conference: E-Medical Systems May 12–14, 2010 –Moracco.
8. M. A. Mohamed El-Bendary, A. E. Abou-El-azm, N. A. El-Fishawy, M. A. R. El-Tokhy, F. Shawki and F. E. Abd El-Samie, "Improving Image Transmission over Mobile Bluetooth Network Using EDR Packets through Proposed

Chaotic Interleaver Design", Accepted Mixdes Conference 2010, 17th International Conference Mixed Design of Integrated Circuits and Systems Wroclaw, 24–26 June 2010.
9. M. A. M. Mohamed, A. Abou El-Azm, N. A. El-Fishawy, M. A. R. El-Tokhy, F. Shawky and F. E. Abd El-Samie "Bluetooth Performance Improvement Over Different Channels Through Channel Coding" Proceedings of the 5th International Multi-Conference on Systems, Signals and Devices, Jordan, 2008.
10. M. A. M. Mohamed, A. Abou El-Azm, N. A. El-Fishawy, M. A. R. El-Tokhy, F. Shawky and F. E. Abd El-Samie "Bluetooth Performance Improvement Using Convolutional Codes" Proceedings of the 2nd International Conference on Electrical Systems Design & Technologies, Hammamet Tunisia, Nov. 8–10, 2008.
11. M. A. M. Mohamed El-Bendary, A. E. Abou-El-azm, N. A. El-Fishawy, F. S. M. Al-Hosarey, M. A. R. El-Tokhy, and F. E. Abd El-Samie,"Optimization of Bluetooth Packet Format for Efficient Performance", Progress in Electromagnetic Research M, Vol. 1, 101–110, 2008.
12. M. A. M. Mohamed El-Bendary, A. E. Abou-El-azm, N. A. El-Fishawy, F. S. M. Al-Hosarey, M. A. R. El-Tokhy, and F. E. Abd El-Samie,"Bluetooth Performance Improvement with Existing Convolutional Codes over AWGN Channel" accepted in 2nd International Conference on Electrical Engineering Design and Technologies (ICEEDT'08) Transaction No. 537 November 8–10, 2008 Hammamet, Tunisia.
13. [M. A. M. Mohamed El-Bendary, A. E. Abou-El-azm, N. A. El-Fishawy, F. S. M. Al-Hosarey, M. A. R. El-Tokhy, and F. E. Abd El-Samie,"Improvement of Bluetooth Performance over Fading Channel through Convolutional Codes", submitted to The 18 International Conferences on Computer Theory and Applications (ICCTA 2008) Alexandria-Egypt.
14. M. A. M. Mohamed El-Bendary, A. E. Abu El-Azm, N. A. El-Fishawy, A. Hagagg, F. Shawy, and F. E. Abd El-Samie,"Effects of Convolutional Encoder Constraint Lengths on Bluetooth Performance over Different Channels", PIERS 2009 Moscow Progress. In Electromagnetics Research Symposium, August 18–21, 2009, Moscow, RUSSIA.
15. M. A. M. Mohamed El-Bendary, A. E. Abu El-Azm, N. A. El-Fishawy, A. Hagagg, F. Shawy, and F. E. Abd El-Samie,"Effects of Different Constraint Lengths of Convolutional Encoder on Bluetooth Performance over Fading Channels", The 19th International Conferences on Computer Theory and Applications, ICCTA 2009, 17–19 October 2009, Alexandria, Egypt.
16. M. A. M. Mohamed El-Bendary, A. E. Abou-El-azm, N. A. El-Fishawy, F. S. M. Al-Hosarey, M. A. R. El-Tokhy, and F. E. Abd El-Samie,"power-efficient Image Transmission over Bluetooth Using Randomization Technique", accepted in 5th International Computer Engineering Conference (ICENCO 2009), December 27–29, 2009, Cairo University, Egypt.
17. M. A. M. Mohamed El-Bendary, A. E. Abou-El-azm, N. A. El-Fishawy, F. S. M. Al-Hosarey, M. A. R. El-Tokhy, and F. E. Abd El-Samie,"Chaotic

Interleaving: A Tool to Enhance the Performance of Bluetooth Systems", Progress In Electromagnetics Research Symposium, Paper Key: 091222143202, PIERS 2010 in Cambridge, USA, 5–8 July, 2010.
18. M. A. M. Mohamed El-Bendary, A. E. Abou-El-azm, N.l A. El-Fishawy, A. Hagagg, F. S. M. Al-Hosarey, and F.i E. Abd El-Samie,"Performance of Real-time Audio Signal Transmission over ACL Link with Channel Interleaving Consideration", Progress In Electromagnetics Research Symposium, Paper Key: 091222142557, PIERS 2010 in Cambridge, USA, 5–8 July, 2010.
19. M. A. M. Mohamed El-Bendary, A. E. Abu El-Azm, N. A. El-Fishawy, A. Hagagg, F. Shawy, and F. E. Abd El-Samie,"EDR Packets Throughput Improving over Bluetooth System for Longer and Shorter Packets through ACQDDR and Adaptive Packets", 6 the International Conference on networked Computing, IEEE INC2010, may 11–13, 2010, Gyeongju, Korea.
20. M. A. M. Mohamed El-Bendary, A. E. Abu El-Azm, N. A. El-Fishawy, A. Haggag, M. A. R. El-Tokhy, F. Shawky, and F. E. Abd El-Samie, "Throughput Improvement over Bluetooth Network through Adaptive Packets", 6 the International Conference on networked Computing, invited paper, IEEE INC2010, may 11–13, 2010, Gyeongju, Korea.

Chapter 1
Introduction

Abstract In this chapter, there are some concepts of Wireless Sensor Networks (WSN) and its parts are presented. The types of sensing station also, are introduced. The WSN and Wireless Body are Network (WBAN) are sensing networks with air medium. These networks shared in many points. The sensing process is the basis of the WSN and WBAN, the main difference between them in the deployment location and the types of applications.

The WSN applications concerns to the environment in general definition. In this book the meaning of environment term is presented. On the other hand, the WBAN application is related to the human body and the closed surround environment.

Keywords Wireless networks • Sensing networks • WBAN • Environment

1.1 Introduction

In this chapter some definition related to the WSN and WBANs are presented. The main job of these types of wireless networks is the sensing or monitoring Changes in the surrounding environment with collecting and processing these changes then make automatic decisions or providing recommendations to the human element. The pretty in this type of networks, it enables new applications and open development of new prospects for application areas wider networks of sensors and more than just be used in the areas of safety and electronic security systems, which is called in the book of traditional sensors networks.

So, the WSN can be defined as a network of devices, denoted as nodes. These nodes can sense the environment and communicate the information gathered from the monitored field (around environment = area or volume) through wireless link. In the next section, the environment definition and its contents have been presented.

The WBAN application concerns the human body health only, the WSN may be utilized for public health environment monitoring. So, with the expected and complementary working the WBANs may be work as a branch of national WSN of public health saving within specific area or governor or complete country.

"Integrated network of public health in a specific area, whether the region in a particular country or the whole of a country" [1].

M.A.M. El-Bendary, *Developing Security Tools of WSN and WBAN Networks Applications*, Lecture Notes in Electrical Engineering 316, DOI 10.1007/978-4-431-55069-3_1

In the WSNs, nodes are simple and low-complexity devices; the typical applications require few bytes sent periodically or upon request or according to some external event; every node can be either source or destination of information, not both; some nodes do not play the role of routers; energy efficiency is very relevant matter, while capacity is not for most application. Therefore, WSN are not a specific case of wireless ad-hoc networks. Thus, a lot of care must be used when considering protocols and algorithms which are good for ad hoc networks, and using them in the context of WSNs.

1.2 Wireless Network Overview

WSN is wireless network oriented to specific mission. This mission is briefed in monitoring something. So, the WSN can be called "Object oriented Wireless Networks", it always in its applications relates to the sensing process. The WSN located within the different types of wireless network, it covers small area and large area. The wide area of research belong this sensing network due to the tremendous advances of wireless networks and its applications [2].

The WSN, This type of network can be considered as covering the spaces between the different types of wireless networks, and even eliminates the geographical differences and challenges among them. Where can this type that works on the physical and personal networks through wireless computer networks and access to wireless networks and wide-ranging. In the following, the different networks are discussed in brief.

Wireless networks are gaining popularity to its peak today, as the user wants wireless connectivity irrespective of their geographic position. Wireless Networks enable users to communicate and transfer data with each other without any wired medium between them. One of the reasons of the popularity of these networks is widely penetration of wireless devices. Wireless applications and devices mainly emphasize on Wireless Local Area Networks (WLANs). This has mainly two modes of operations, i.e. in the presence of Control Module (CM) also known as Base Stations and Ad-Hoc connectivity where there is no Control Module. Ad-Hoc networks do not depend on fixed infrastructure in order to carry out their operations. The operation mode of such network is stand alone, or may be attached with one or multiple points to provide internet and connectivity to cellular networks [3].

These networks exhibits the same conventional problems of wireless communications i.e. bandwidth limitations, battery power, enhancement of transmission quality and coverage problems.

1.2.1 Network Concepts

Before going into the details of wireless network, it is important to understand what a network is and different kind of networks available today. Any collection of devices/computers connected with each other by means of communication channels that help the users to share resources and communicate with other users. There are two main types of network i.e. wired network and wireless network. So, with merging the sensor term within the network term definition, the meaning of "sensor network" becomes clear and meaningful [4].

1.2.2 Wired Networks

Wired network are those network in which computer devices attached with each with help of wire. The wire is used as medium of communication for transmitting data from one point of the network to other point of the network.

1.2.3 WSN Real-Time Network

So-called sensor nodes combine means for sensing environmental parameters, processors, wireless communication capabilities, and autonomous power supply in a single compact device. Networks of these untethered devices can be deployed unobtrusively in the physical environment in order to monitor a wide variety of real-world phenomena with unprecedented quality and scale while only marginally disturbing the observed physical processes [5].

1.2.4 Traditional Sensing Network

The backbone of WSN and WBAN networking applications are based on the first process in the other side of these system which is the sensing element "Sensing process = sensor". The idea of objects monitoring or sensing of the surround environment is not novel idea. The most common wired sensors network is the fire alarm cable network. Table 1.1 gives simple comparison between the features of traditional networks and the wireless sensor networks.

Table 1.1 Traditional network and wireless sensor network comparison

Traditional networks	Wireless sensor networks
General purpose design; serving many application	Single purpose design; serving one specific applications
Typical primary design concerns are network performance and latencies; energy is not a primary concern	Energy is the main constraint in the design of all node and network components
Networks are designed and engineered according to plans	Deployment, network structure, and resource use are often ad hoc (without planning)
Devices and networks operate in controlled and mild environments	Sensor networks often operate in environments with harsh conditions
Maintenance and repair are common and networks are typically easy to access	Physical access to sensor nodes is often difficult or even impossible
Component failure is addressed through maintenance and repair	Component failure is expected and addressed in the design of the network
Obtaining global network knowledge is typically feasible and centralized management is possible	Most decisions are made localized without the support of a central manager

1.2.5 Remote and Wired Sensing

Many existing systems for observing real-world phenomena are based on few sensors with a relatively long range, such as satellites for earth observation, weather stations, or sonar. Due to the long range, these systems can observe phenomena that are far away from the sensors. However, the resolution of these systems decreases with the observation distance. Moreover, many systems require a free line of sight between the sensors and the observed phenomenon.

Wireless sensor nodes are equipped with short range sensors. Many of these devices are placed in the close vicinity of the observed phenomenon. Since the average distance between the observed phenomenon and the sensors is small and since many redundant sensors observe a single phenomenon, the effective monitoring resolution of a wireless sensor network can be better than that of a remote sensing approach. The placement of the sensors close the phenomenon allows the use of wireless sensor networks also in cluttered environments where line-of-sight paths are rather short.

Many existing systems make use of distributed, wired sensors that are connected to a central computing device (e.g., sensors in cars and engines, sonar arrays). Such an approach has a number of advantages: sensors do not need separate power supplies, the wired network has a fixed topology, small and deterministic delays, and communication errors are very rare. On the other hand, the wiring limits the flexibility and scale of such wired sensor networks.

1.2.6 Different Wireless Networks

A network in which, computer devices communicates with each other without any wire. The communication medium between the computer devices is wireless. When a computer device wants to communicate with another device, the destination device must lays within the radio range of each other. Users in wireless networks transmit and receive data using electromagnetic waves. Recently wireless networks are getting more and more popular because of its mobility, simplicity and very affordable and cost saving installation [6].

Why Wireless Networks?
Wireless networks are getting popular due to their ease of use. Consumer/user is no more dependent on wires where he/she is, easy to move and enjoy being connected to the network. One of the great features of wireless network that makes it fascinating and distinguishable amongst the traditional wired networks is mobility. This feature gives user the ability to move freely, while being connected to the network. Wireless networks comparatively easy to install then wired network. There is nothing to worry about pulling the cables/wires in wall and ceilings. Wireless networks can be configured according to the need of the users. These can range from small number of users to large full infrastructure networks where the number of users is in thousands.

Wireless networks are very useful for areas where the wire cannot be installed like hilly areas.

On the basis of coverage area the wireless network can be divided into:

(a) Body Area Network
(b) Personal Area network
(c) Local Area Network
(d) Wide Area Network

The WSN means that there are two concepts it should be known, sensing something and monitoring some objects or something. So the goal or the WSN networks functionality is transferring the changes in the surround environment to be monitored and controlling its predicted results. The term of environment understanding is very useful to know how utilize successfully the WSN. The applications of WSN are related mostly to the environment term. As shown in Fig. 1.1, there is similarity between the different environments as shown in the figure and the arrangement of different wireless networks ranges. The WBAN is wireless sensing network locally working; it is one of wireless networks as shown in figure. The WSN is sensing network also, but it can be considered a globally working net. There are many definition of the environment term are discussed in Chap. 2. In brief, the environment can be defined in the following [7, 8].

Physical Environment
Environment is both physical and biological. It includes both living and non-living components.

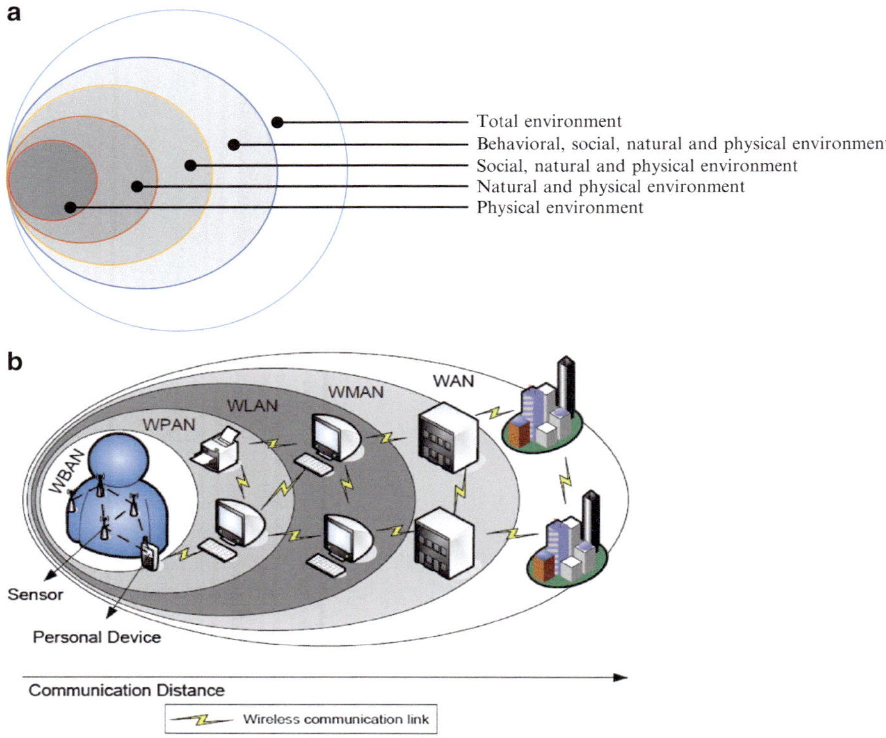

Fig. 1.1 The environment and wireless technologies similarity. (**a**) The environment sections structure. (**b**) The different wireless networks technologies

"The Physical Environment is classified into three broad categories viz. (*i*) Solid, (*ii*) Liquid (*iii*) Gas. Solid environment and its WSN applications, such as monitoring and predicting occurrence or the probability of occurrence of earthquakes and volcanoes or earthquakes (ground crust cracks). In order to minimize the losses that may result from them".

Biological Environment
The biological of the environment consists of: (i) Plants, (ii) Animals. The human and animals are greatly influenced by the climate and the weather, through Predict weather changing. Liquid and includes monitor and predict hurricanes, tidal shifts and fleecing of the seas and oceans.

Definition Environment literally means surrounding and everything that affect an organism during its lifetime is collectively known as its environment.

The expected WSN advanced application can be provided the capability to control (to improve and to choose) environmental health concerns both living and working environments for individuals, social groups and communities and marks the pathway toward equity in health as a goal of the human development. In this framework, poverty and deprivation are also measured in terms of vulnerability to control living in health and in healthy environment and concern functioning related to environmental health for individuals, social groups and communities. Therefore, environmental health can be seen as a social condition embedded in public health, epidemiological environment and public development policies; their interconnections strongly affect the relation between individual income and capability to be healthy within a society [9].

WBAN can be classified to human body medical care service, so, the surrounded environment may be monitored by closed WSN to decrease or prevent any harm to human body. So, *the environment is all of physical and social conditions that surround a person and can influence that person's health. Your* ***physical environment*** *includes both your outdoor and indoor surroundings. The quality of air you breathe and the water you drink are important to your health. So is your exposure to deceasing- causing orgasms, to loud noise, and to radiation from the sun and other sources.*

- ***The WSN and WBAN can be integrated within the human care application through integrated care system to serve the public health***.

Environmental health is that branch of public health that is concerned with all aspects of the natural and built environment that may affect human health. Other phrases that concern or refer to the discipline of **environmental health** include **environmental public health**, and **environmental protection**. The field of environmental health is closely related to environmental science and public health as environmental health is concerned with environmental factors affecting human health.

1.3 Sensing Networks

The book focuses on the two familiar wireless sensing networks which are, the WSN and WBAN. There are many shared lines between WSN and WBAN networks. The idea of sensing network is not a new subject. There is traditional or wired sensing network such as the fire alarm network. This network is based on the sensing process like the WSN idea; the main appeared difference is the type of media. The fire alarm network is established for detect the fire happening using different types of detectors, such as the flame detectors, smoke or heat detectors. There are conventional fire alarm systems, and intelligent system. For example, the forest fire detection using WSN application can be considered the fire alarm network similar the traditional fire alarm networks but outdoor within huge area [10].

In Fig. 1.2 the traditional wired sensing network (Fire alarm system) and the WSN. As shown in this figure, the Fire Alarm Control Panel (FACP) acts as sink "controller" in the WSN configuration. Also, the sensors are the sensing station "nodes" for any types such as smoke detectors, heat detectors, or flame detectors.

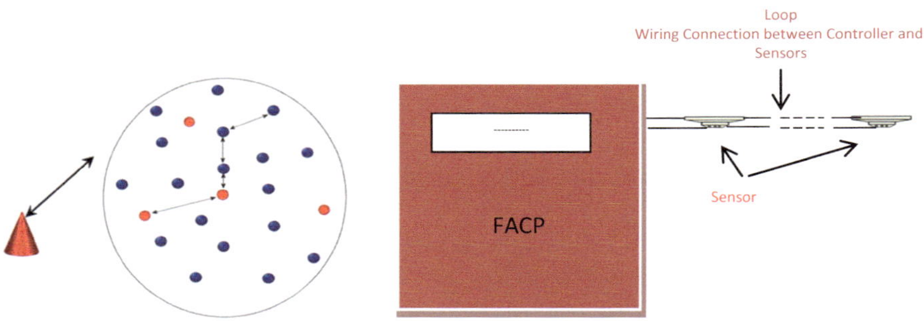

Fig. 1.2 Traditional sensing network and WSN simple contents. (**a**) WSN contents. (**b**) Traditional sensing wired network

Fig. 1.3 The sensing node contents

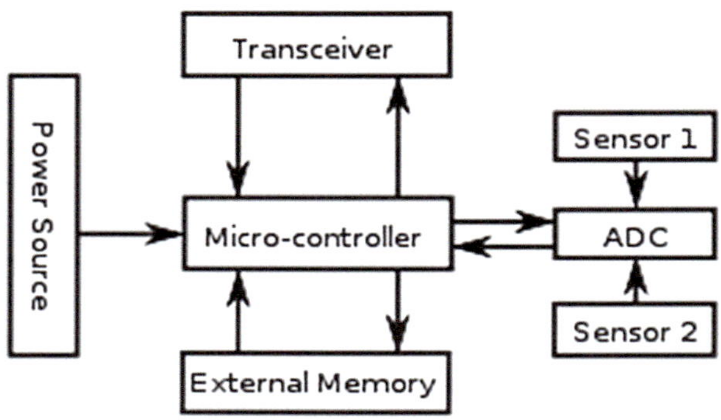

This example is devoted for implement the WSN idea by familiar example such as the wired sensing network of fire alarm system [11].

The WSN is wireless network contains number of devices called nodes. The node cab be called the Sensing Station (SS) or Sensing Node (SN), it can be monitor or detect the changes in the specific area "service area" and send the captured event to the controller "sink". Figure 1.3 shows the contents of the sensing station "Sensing Node" (SN).

So, the WSN contains the controller unit "Sink" and number of SN. In this book, there are many types of SN will be discussed, the conventional and fixed nodes, the mobile nodes and the mobile and intelligent nodes. Also, the classical and cooperative WSN are discussed in this book [12].

Ad-hoc network is defined as a wireless network without infrastructure. This simple definition compared to some of WSN application is very suitable and closed to the idea of sensing station deploying. The WBAN is the same or not? May it can be seen the single body as one sensing station and it in mobile and continuous moving, but mostly the sensors which are monitoring the bio-function of the human body is in fixed position. *So, the sensing station in WSN and WBAN shared some properties and different in another*.

1.4 WSN Overview

A **wireless sensor network** (WSN) is a wireless network consisting of spatially distributed autonomous devices using sensors to cooperatively monitor physical or environmental conditions, such as temperature, sound, vibration, pressure, motion or pollutants, at different locations. The development of wireless sensor networks was originally motivated by military applications such as battlefield surveillance. However, wireless sensor networks are now used in many civilian application areas, including environment and habitat monitoring, healthcare applications, home automation, and traffic control [13].

WSN is the wireless network which is related to communicate with the surrounding medium. Almost its application belong directly or indirectly the environmental health. One environmental monitoring application of WSN is the air and water pollution monitoring; these two elements are considered the main of environment elements. As shown in Fig. 1.4, there are many ways to deploying the Sensing Stations (SN), nodes within the service volume area.

"It is a wireless networks have arms to sense and have eyes to see and monitor the different natural phenomena and sense the changes of the objects in the surround environment. The arms and eyes denotes to the sensors of the sensing station nodes of WSN. These networks provide us the mirror to see what indirectly seen."

Sensors definition:
The node of WSN is a device from many devices in the network, it denotes to the part of sensing process and as well as the associated operations which follows the sense process such as the signal processing, amplifications, ADC… and so on. A **sensor node**, also known as a **mote** "effective small part", it is can be defined as a node in a wireless sensor network that is capable of performing some processing, gathering sensory information and communicating with other connected nodes in the network. A mote is a node but a node is not always a mote.

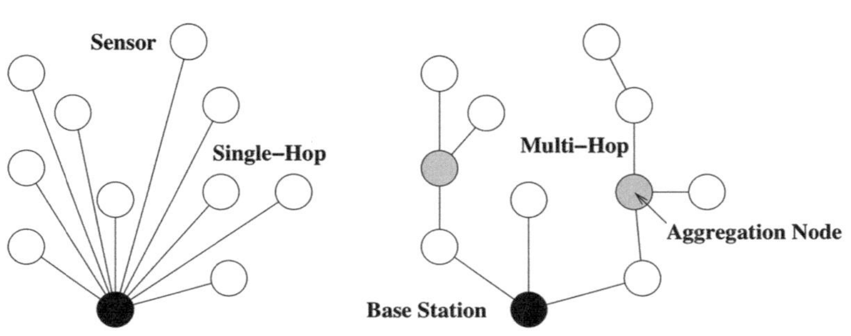

Fig. 1.4 Single-hop versus multi-hop communication in sensor networks

The role of the sink in WSN:
The controller performs tasks, processes data and controls the functionality of other components in the sensor node. While the most common controller is a microcontroller, other alternatives that can be used as a controller are: a general purpose desktop microprocessor, digital signal processors, FPGAs and ASICs. A microcontroller is often used in many embedded systems such as sensor nodes because of its low cost, flexibility to connect to other devices, ease of programming, and low power consumption. A general purpose microprocessor generally has higher power consumption than a microcontroller; therefore it is often not considered a suitable choice for a sensor node. Digital Signal Processors may be chosen for broadband wireless communication applications, but in Wireless Sensor Networks the wireless communication is often modest: simpler, easier to process modulation and the signal processing tasks of actual sensing of data is less complicated. Therefore the advantages of DSPs are not usually of much importance to wireless sensor nodes. FPGAs can be reprogrammed and reconfigured according to requirements, but this takes more time and energy than desired [14].

1.5 WBAN Overview

Figure 1.5 shows one of WPAN applications scenario for medical services. It is cleared from this figure, the WPANS medical applications role is not stopped at the medical functions, it also, involves non medical role as shown in the left of Fig. 1.5.

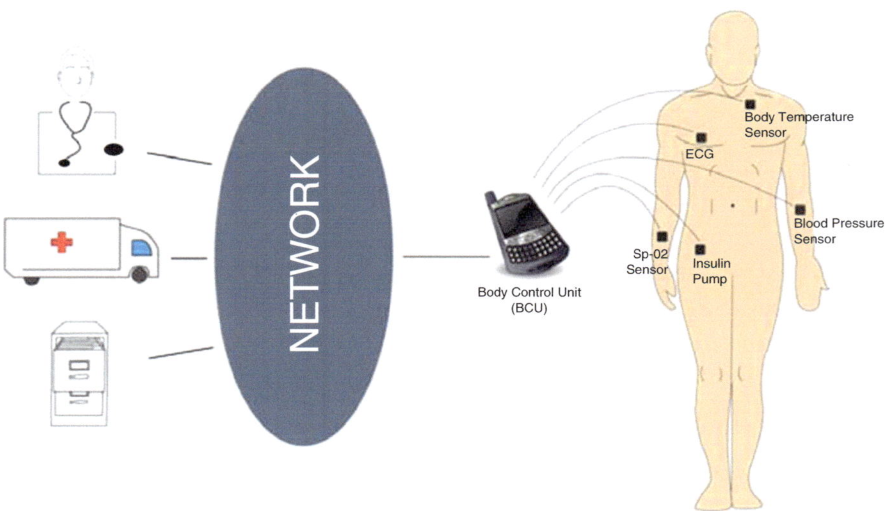

Fig. 1.5 Example scenario of WBAN applications

Wireless network is its primary purpose medical applications It aims to provide care and follow-up of patients and the possibility of making patient monitoring around the clock and move the vital signs of the human body to the hospital or medical center for wireless connections, whether the patient is fixed or moving near or far from the center of medical service.

It also aims at applications of these networks to reduce the risk of critical situations for patients with health conditions such as heart patients and sugar and through constant contact, which can transfer any change monitored by sensors biomarkers found on the body of the patient to a specialist doctor to take too much about what actions are necessary to maintain the patient's life.

Is a set of devices to be named sensation or inside the patient's body is connected wirelessly with each other and with the controller assigned to it to receive signals from the sensors.

WBAN refers to medical applications of wireless networks; it can be called a health care wireless technology. A body area network (BAN), also referred to as a wireless body area network (WBAN) or a body sensor network (BSN), is a wireless network of wearable computing devices [15–17]. In particular, the network consists of several miniaturized body sensor units (BSUs) together with a single body central unit (BCU) [18, 19]. The development of WBAN technology started around 1995 around the idea of using wireless personal area network (WPAN) technologies to implement communications on, near, and around the human body. About 6 years later, the term "BAN" came to refer systems where communication is entirely within, on, and in the immediate proximity of a human body [20, 21]. A WBAN system can use WPAN wireless technologies as gateways to reach longer ranges.

A typical wireless body area network (WBAN) comprises of a number of wireless sensor nodes and a body gateway. It is a distributed system with various hardware configurations and operating systems. The middleware layer bridges between applications and the underlying networking/RTOS protocol stack so as to simplify the application development, and make codes re-usable from one application to another.

The rapid growth in physiological sensors, low-power integrated circuits, and wireless communication has enabled a new generation of wireless sensor networks, now used for purposes such as monitoring traffic, crops, infrastructure, and health. The body area network field is an interdisciplinary area which could allow inexpensive and continuous health monitoring with real-time updates of medical records through the Internet. A number of intelligent physiological sensors can be integrated into a wearable wireless body area network, which can be used for computer-assisted rehabilitation or early detection of medical conditions. This area relies on the feasibility of implanting very small biosensors inside the human body that are comfortable and that don't impair normal activities. The implanted sensors in the human body will collect various physiological changes in order to monitor the patient's health status no matter their location. The information will be transmitted wirelessly to an external processing unit. This device will instantly transmit all information in real time to the doctors throughout the world. If an emergency is detected, the physicians will immediately inform the patient through the computer system by sending appropriate messages or alarms. Currently the level of information provided and energy resources capable of powering the sensors are limiting. While the technology is still in its primitive stage it is being widely researched and once adopted, is expected to be a breakthrough invention in healthcare, leading to concepts like telemedicine and m Health becoming real.

Applications
Initial applications of BANs are expected to appear primarily in the healthcare domain, especially for continuous monitoring and logging vital parameters of patients suffering from chronic diseases such as diabetes, asthma and heart attacks.

- ***A BAN network in place on a patient can alert the hospital, even before they have a heart attack, through measuring changes in their vital signs.***
- ***A BAN network on a diabetic patient could auto inject insulin through a pump, as soon as their insulin level declines.***

Other applications of this technology include sports, military, or security. Extending the technology to new areas could also assist communication by seamless exchanges of information between individuals, or between individual and machines.

1.6 WSN and Embedded System Relation

An embedded system consists of hardware and software components, where the allocation of functions to hardware and software is highly dependent on the requirements of the actual system. For performance and security reasons, part of the functionality may be provided by application-specific hardware components. The use of software in conjunction with programmable hardware provides flexibility and support for more complex features.

What differentiates wireless sensor networks from traditional embedded systems is the large number of participating nodes, the use of wireless communication, and often severe resource/cost/energy constraints of sensor nodes. Traditional embedded systems are typically a fixed part of highly engineered structures, that is, the actual "embedding" takes place at production time and do not change during the lifetime of the embedded system. In contrast, wireless sensor nodes may be deployed in a natural setting with little control over the actual placement and distribution of the nodes. Also, the embedding (i.e., deployment) of sensor nodes into the physical environment is not tied to the production time of the sensor node and may change during the lifetime of a sensor node [22].

The "Ad-hoc" is Latin phrase; its (*linguistically*) semantically meaning is "***for purpose***".

The scientific definition of Ad-hoc, it is the wireless network contains a group of nodes, self-organized, self configured. The WSN is wireless network, it is established and built for pre-determined function and known purpose.

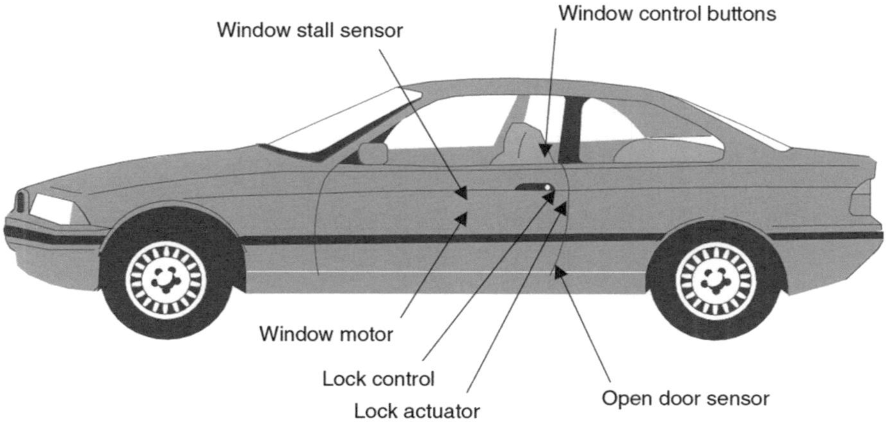

Fig. 1.6 Embedding systems within car for example

The Ad-hoc network examples is (Industry consortia, such as Bluetooth and home RF are already designing standard for wireless networks of embedded devices that can serve as communication backbone of sensor networks.) WSN is a real time embedded wireless system.

Embedded system can be defined as a system whose principal function is not computational, but which is controlled by a computer embedded within it. Figure 1.6, for example, it is presented as applicable embedded system example. The embedded system appears by its effects and its function execution. As shown in Fig. 1.6, there are many embedded systems in this car make specific function, but in the reality these systems are not seen by the user, he senses it by its effects.

Def A microprocessor-based control system which processes a fixed set of programmed instructions to control electromechanical equipment which may be part of an even larger system. It refers to either single or multi-purpose computerized devices that are literally embedded within some larger piece of engineering equipment or industrial product [23].

WPANs definition

A wireless personal area network (WPAN) is a personal area network – a network for interconnecting devices centered around an individual person's workspace – in which the connections are wireless. Wireless PAN is based on the standard IEEE 802.15. The two kinds of wireless technologies used for WPAN are Bluetooth and Infrared Data Association.

A WPAN could serve to interconnect all the ordinary computing and communicating devices that many people have on their desk or carry with them today; or it could serve a more specialized purpose such as allowing the surgeon and other team members to communicate during an operation.

A key concept in WPAN technology is known as "plugging in". In the ideal scenario, when any two WPAN-equipped devices come into close proximity (within several meters of each other) or within a few kilometers of a central server, they can communicate as if connected by a cable. Another important feature is the ability of each device to lock out other devices selectively, preventing needless interference or unauthorized access to information.

1.7 Wireless Sensing Networks Contents

There are many important and main terms format the wireless sensing networks concepts. In this book, these terms are described in role every one of them. The terms as follows:

WSN, it is a group of devices (nodes), these devices are wireless connected together. It forms the "Sensing Network".
WBAN, it is Medical Application Sensing network.
Sensors and Detectors, it is the sensing part in the node. It is tiny "mote" device is installed within the node "one of the node contents".
Coverage area, it is the maximum detection or sensing detector or sensor, section from the service area.
Head, it is the boss of number of nodes communicates them to the sink.
Cluster, it is a group of nodes service specific area as section from the whole service area, and contains number of coverage area.
Service area, it is the volume which must be monitored by the WSN.
Sink, it is the brain of WSN; it may be the decision maker or act as a gateway only.

The wireless sensing network, it is a group of number of nodes cooperatively work to monitor specific area. The book presents cooperative nodes within the wireless sensing network. This idea presents deeply cooperation concept more than the understood from the word "cooperatively working" [24].

- Multiple sensors (often hundreds or thousands) form a network to ***"cooperatively monitor large or complex physical environments. The book discusses the proposed WSN – cooperative nodes based on the secret key algorithm are used to define the adjacent nodes within the cluster area or the whole service area".***
- Acquired information is wirelessly communicated to a base station (BS), which propagates the information to remote devices for storage, analysis, and processing.

In the same manner, the WBAN, it contains number of sensing stations to sense the human bio-effects of the human body. The coverage area in the WBAN is specific area in the human body or human organ such as the heart. The sensors of WBAN, senses the human body signs only, heart, brain, blood pressure, and so on. Every one of them needs enough number of sensors. The previous all terms many are contained in the sensing network or the network miss number of them. For example, some applications scenarios do not need and the cluster. As shown in Fig. 1.1, the sensing stations connected directly to the sink. Also, the WBAN miss number of these terms such as the head is needed. The cluster concept with WBAN application is not suitable. The service area and the coverage area almost is part in the human body.

The nodes can be classified according to its features as follows types: "***Classical node, intelligent node, Mobile node and intelligent mobile node***"**.**

Function of intelligent nodes in the mobile WSN is discussed based on the security algorithm.

The cooperative WSN past work is focused on the routing and localization algorithms. The researcher looked at the cooperation concept is investigated already by default in the WSN operation. This view is true within the multi-hop WSN Topology or all nodes serve the same volume and monitor the same object so, they in the cooperative work. But in the truth, if there is no cooperation between the nodes together, there is misunderstanding in the cooperation idea. It is not cooperative WSN.

"It aims to let the SN can talk to adjacent SN to confirm or deny any detected event or signal, the Mobile SN (MSN) can walk to be near another SN or move away from the SN according to the captured event or the environmental changes" it is the means of Cooperative WSN and intelligent mobile SN.

The WSN suffers from many challenges; the book presents many scenarios with the simulation results and Matlab codes to decrease the complexity of the communications system. The presented scenarios can be applied for WSN applications.

This work focused on the mobility options of the WSN nodes sensing stations. The mobility capability of the sensing station has several advantages, such as enhances the flexibility deployment of the nodes changing. The mobility of the nodes with utilizing cooperation algorithm between the SN, this leads to decreasing the number of nodes to cover specific service area. One of the desirable aspects of wireless sensor nodes is communication between them the wireless link. That leads to, mobile applications can be supported; flexible deployment of nodes is possible; and the nodes can be placed in areas that are otherwise inaccessible to wired nodes. Once the deployment is carried out, it is possible to rearrange node placement in order to attain optimal coverage and connectivity; and the rearrangement can be made without disrupting the normal operation of the structure or process the nodes monitor [25].

The security issue is essential factor in the WSN application. There are many important application of WSN such as the military field and others. Let us suppose the WSN is utilized to detect the nuclear radiation surround nuclear site and the sensing stations gives the confirmed alarm, the governmental authorities will cost high price for nothing. Finally, some of WSN application may be related to the national security of the counties. In this book, the multi-level security techniques are presented [26].

The development of WSN technologies is rapidly growth. Soon, the mobile WSN networks will be implemented. This point is covered in the book. The mobility effects on the transmitted signal and the bad effects combating through different techniques with the Matlab codes of the different scenarios are presented.

1.8 Wireless, Mobile, and Ad Hoc Networks

Wireless communication, especially with focus on short communication range and low power consumption, is a key enabling technology for wireless sensor networks. In mobile networks, computers capable of wireless communication can change their physical position over time, resulting in dynamically changing network topologies. Ad hoc networks are wireless networks that do not require an external infrastructure such as base stations in mobile phone networks. The nodes of an ad hoc network act both as sources/sinks of messages and as routers that forward messages on behalf of other nodes. Nodes can join and leave the network anytime. Although ad hoc networks may also consist of immobile nodes, they often contain mobile nodes. Power awareness is an important issue in the context of mobile networks, since mobile computing devices are often powered by batteries. Recent research in mobile ad hoc networks focuses on routing, mobility management, power management, self-configuration, and the radio interface (including the radio hardware and medium access techniques) [27].

It is anticipated that many wireless sensor networks will be implemented as a mobile ad hoc network (MANET). However, results from MANET research often cannot be directly applied to wireless sensor networks, since resource and energy constraints are typically more stringent here. Typical MANET research focuses on handheld devices or laptops with renewable batteries. The computing, storage, communication resources of these devices are comparable to desktop computers.

In contrast, sensor node batteries are often not replaceable; range, bandwidth, reliability of wireless communication links, computing and memory resources, and available energy may be orders of magnitude smaller compared to more traditional MANET nodes [28].

Wireless sensor networks may also rely on infrastructure-based mobile networks. For example, mobile phone companies are currently exploring the value of mobile phones for sensor networks. Such networks could either solely consist of mobile phones equipped with sensors, or a mobile phone could act as a gateway connecting an ad hoc sensor network to the phone network. Such combinations of infrastructure-based and ad hoc networks would allow remote access to sensor networks and integration with existing computing infrastructures. In Fig. 1.7, the nodes A, B, C and D are considered as sensing nodes [29].

A **wireless ad hoc network** is a decentralized type of wireless network. The network is ad hoc because it does not rely on a pre existing infrastructure, such as routers in wired networks or access points in managed (infrastructure) wireless networks. Instead, each node participates in routing by forwarding data for other

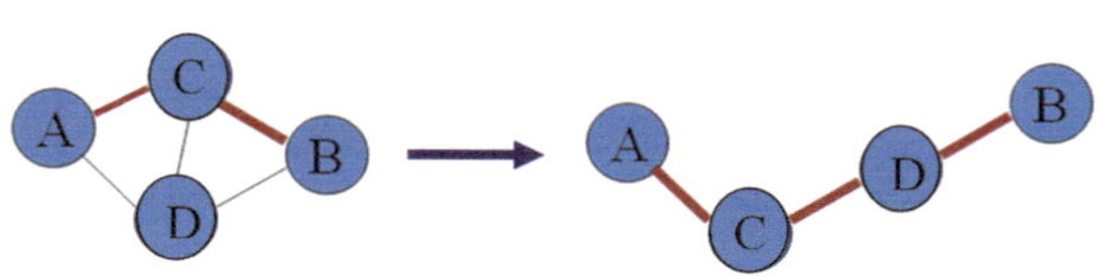

Fig. 1.7 WSN as wireless ad hoc network

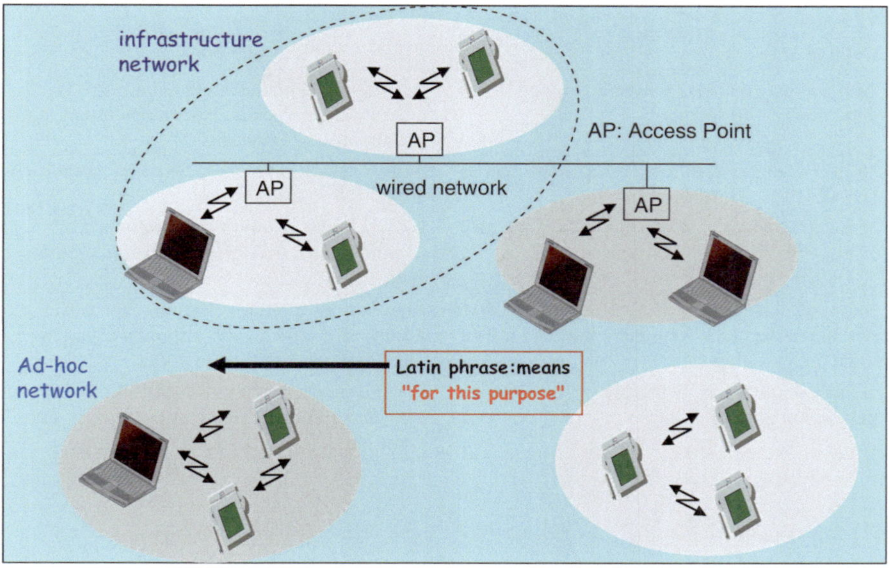

Fig. 1.8 Infrastructure of wired and ad-hoc networks

nodes, so the determination of which nodes forward data is made dynamically on the basis of network connectivity. In addition to the classic routing, ad hoc networks can use flooding for forwarding the data.

An ad hoc network typically refers to any set of networks where all devices have equal status on a network and are free to associate with any other ad hoc network device in link range. Ad hoc network often refers to a mode of operation of IEEE 802.11 wireless networks as shown in Fig. 1.8.

It also refers to a network device's ability to maintain link status information for any number of devices in a 1-link (aka "hop") range, and thus, this is most often a Layer 2 activity. Because this is only a Layer 2 activity, ad hoc networks alone may not support a routable IP network environment without additional Layer 2 or Layer 3 capabilities [30].

The decentralized nature of wireless ad hoc networks makes them suitable for a variety of applications where central nodes can't be relied on and may improve the scalability of networks compared to wireless managed networks, though theoretical and practical limits to the overall capacity of such networks have been identified.

Minimal configuration and quick deployment make ad hoc networks suitable for emergency situations like natural disasters or military conflicts. The presence of dynamic and adaptive routing protocols enables ad hoc networks to be formed quickly. Wireless ad hoc networks can be further classified by their application. There is main difference between the cellular networks and the Ad-hoc wireless networks in the applications and the network establishing. As shown in Table 1.2, it gives brief comparison between these two wireless systems.

Table 1.2 Cellular networks and ad-hoc networks comparison

Cellular networks	Ad Hoc wireless networks
Application domains include mainly civilian and commercial sectors	Application domains include battlefields, emergency search and rescue operations, and collaborative computing
High cost of network maintenance (backup power source, staffing, etc.)	Self-organization and maintenance properties are built into the network
Mobile hosts are of relatively low complexity	Mobile hosts require more intelligence (should have a transceiver as well as routing/switching capability)
Major goals of routing and call admission are to maximize the call acceptance ratio and minimize the call drop ratio	Main aim of routing is to find paths with minimum overhead and also quick reconfiguration of broken paths
Widely deployed and currently in the third generation of evolution	Several issues are to be addressed for successful commercial deployment even though widespread use exists in defense

- ***mobile ad hoc networks*** ***(MANET)***

The previous section means the Ad-Hoc concepts is not abdicable for WSN applications, if the node in the WSN sends its data to the sink; it is not Ad-Hoc network. WSN should have infrastructure to complete its task.

Ad hoc networks are a new paradigm of wireless communication for mobile hosts (which we call nodes). In an ad hoc network, there is no fixed infrastructure such as base stations or mobile switching centers. Mobile nodes that are within each other's radio range communicate directly via wireless links, while those that are far apart rely on other nodes to relay messages as routers. Node mobility in an ad hoc network causes frequent changes of the network topology. Figure 1.7 shows such an example: initially, nodes A and D have a direct link between them. Since an ad hoc network can be deployed rapidly with relatively low cost, it becomes an attractive option for commercial uses such as sensor networks or virtual classrooms [31].

In brief, it should be noted that:

- Wireless Sensor Networks can be considered as a special category of ad hoc networks that are used to provide a wireless communication infrastructure among the sensors deployed in a specific application domain.
- A sensor network is a collection of a large number of sensor nodes that are deployed in a particular region [32].
- Distinct properties of wireless sensor networks:
  - Mobility of nodes is not needed in all cases in wireless sensor networks without cooperative algorithm between the nodes. This point will be explained in Chap. 6.
  - The size of the network is much larger than that in a typical ad hoc wireless network.

- The density of nodes in a sensor network varies with the domain of application.
- In WBAN the controller "sink" is not the decider, the sink of WSN in some scenarios works as a decider.
- WSN may be work locally or globally application, WBAN works within limited area to provide specific service [33].

1.9 Book Organizing

Chapter 2 presents the wireless networks types. The short range technologies such as the Bluetooth and ZigBee technologies are discussed in details. The BT and ZB technologies are the WPAN networks; also, they are utilized to implement the WSN application.

Chapter 3, the sensing networks concepts are presented. The WSN types based on applications types, Event Detection (ED) and Spatial Process Estimation (SPE) are presented. The shared lines between the WSN and WBAN are discussed. Also, the chapter presents the environment meaning and the similarity between the environment grading and the wireless technologies. The wireless networks starts from the WBANs to the WMAN and the location of WSN in the wireless networks arranging. The different application scenarios of WSN and WBAN are discussed.

Chapter 4, the performance analysing of the WSN is presented and the throughput concepts of WSN with the topology variety. The area throughput is a true term or not for the success transmitted data. No doubt, the role of error control schemes in the wireless communications systems is important. It became one of the basic contains of the wireless communications, with the two form the error detection only and error detection with correction form. In general, the main purpose of a channel encoder is to produce a sequence of data that is robust to noise and to provide error detection and forward error correction mechanisms. In simple and cheap transceivers, forward error correction is costly and, therefore, the task of channel encoding is limited to the detection of errors in packet transmission.

Chapter 5, this chapter presents the embedded security capability using effective interleaving techniques. The presented interleavers are based on the chaotic system concepts. The cooperative WSN is discussed using the chaotic techniques. The complexity analysis of the proposed scenarios also is described. The mobility combating is considered also.

References

1. Astrin W, Li H-B, Kohno R (2009) Standardization for body area networks. IEICE Trans Commun E92.B(2):366–372
2. Chen W-Z, Ou W-W, Lu T-Y, Chou S-T, Yang S-Y (2010) A 2.4 GHz reference-less wireless receiver for 1 Mbps QPSK demodulation. In: Proceedings of the IEEE international symposium on circuits systems, pp 1627–1630

3. Maharaj S. Introduction into wireless technology: WLANs, WMANs and WPANs. http://www.ee.ukzn.ac.za/research/
4. Saleem S, Ullah S, Kwak KS (2011) A study of IEEE 802.15.4 security framework for wireless body area networks. Sensors 11(2):1383–1395
5. Lozano C, Rodriguez O (2007) Design of forest fire early detection system using wireless sensor networks. Online J Electron Electr Eng (OJEEE) 3(2):402–405
6. IEEE P802.15 Study Group Medical Body Area Networks (SGMBAN). SG BAN Montreal Interim Meeting Minutes, 20 May 2007
7. IEEE P802.15 Study Group Medical Body Area Networks (SGMBAN). SG BAN San Francisco Plenary Meeting Minutes, 16 August 2007
8. Khedo KK, Perseedoss P, Mungur A (2010) A wireless network sensor air pollution monitoring system. IJWMN 2(2):31–45
9. Garzon CAL, Riveros OJR (2010) Temperature, humidity and luminescence monitoring system using wireless sensor networks (WSN) in flowers growing. In: ANDESCON, 2010 IEEE, pp 1–4
10. Akyilidiz IF, Su W, Sankarasubramaniam Y, Cayirci E (2002) IEEE Communications Magazine, August 2002. Georgia Institute of Technology. A Survey on Sensor Networks, IEEE Communications Magazine 40(8):102–114
11. Drago S, Leenaerts DMW, Sebastiano F, Breems LJ, Makinwa KAA, Nauta B (2010) A 2.4 GHz 830 pJ/bit duty-cycled wake-up receiver with −82 dBm sensitivity for crystal-less wireless sensor nodes. In: Proceedings of the IEEE international solid-state circuits conference, digest of technical papers, pp 224–225
12. Estrin D, Govindan R, Heidemann J, Kumar S (1999) Next century challenges: scalable coordination in sensor networks. In: Proceedings of the fifth annual international conference on mobile computing and networks (MobiCOM '99), Seattle
13. Estrin D, Girod L, Pottie G, Srivastava M (2001) Instrumenting the world with wireless sensor networks. In: Proceedings of the international conference on acoustics, speech and signal processing (ICASSP 2001), Salt Lake City
14. Zulhani Rasin, Hizzi Hamzah, Mohd Shahrieel Mohd Aras (2009) Application and evaluation of high power Zigbee based wireless sensor network in water irrigation control monitoring system. In: IEEE symposium on industrial electronics and applications (ISIEA 2009), 4–6 Oct 2009, Kuala Lumpur
15. Xu Y, Heidemann J, Estrin D (2001) Geography-informed energy conservation for ad hoc routing. In: Proceedings of the Mobicom, pp 70–84
16. Sinha A, Chandrakasan A (2001) Dynamic power management in wireless sensor networks. IEEE Design Test Comput 18(2):62–74
17. Hughes L, Wang X, Chen T (2012) A review of protocol implementations and energy efficient cross-layer design for wireless body area networks. Sensors 12:14730–14773. doi:10.3390/s121114730
18. Sung W-H, Jui-Yuan Y, Lee C-Y (2011) A robust frequency tracking loop for energy-efficient crystalless WBAN systems. IEEE Trans Circuit Syst II Exp Brief 58(10):637–641
19. Body Area Networks (BAN) (2007) IEEE 802.15 WPAN Task Group 6 [Online]. http://www.ieee802.org/15/pub/TG6.html
20. Wong C-W, McDonagh D, Kathiresan G, Omeni OC, El-Jamaly O, Chan TC-K, Paddan P, Burdett AJ (2008) A 1 V, micropower system-on-chip for vital-sign monitoring in wireless body sensor networks. In: Proceedings of the IEEE international solid-state circuits conference, digest of technical papers, pp 138–139
21. Lhermet H, Condemine C, Plissonnier M, Salot R, Audebert P, Rosset M (2007) Efficient power management circuit: thermal energy harvesting to above-IC microbattery energy storage. In: Proceedings of the IEEE international solid-state circuits conference, digest of technical papers, pp 62–63
22. Wener-Allen G, Johnson J, Ruiz M, Lees J, Welsh M (2005) Monitoring volcanic eruptions with a wireless sensor networks. In: EWSN 2005, Istanbul

23. Whitehouse K, Culler D (2002) Calibration as parameter estimation in sensor networks. In: WSNA, Atlanta
24. Xu N, Rangwala S, Chintalapudi KK, Ganesan D, Broad A, Govindan R, Estrin D (2004) A wireless sensor network for structural monitoring. In: Sensys 2004, Baltimore
25. Ye W, Heidemann J, Estrin D (2002) An energy-efficient MAC protocol for wireless sensor networks. In: IEEE Infocom 2002, New York
26. Dargie W, Poellabauer C (2010) Fundamentals of wireless sensor networks theory and practice. John Wiley and Sons Ltd, United Kingdom.
27. Calhoun BH, Daly DC, Verma N, Finchelstein DF, Wentzloff DD, Wang A, Cho SH, Chandrakasan AP (2005) Design considerations for ultralow energy wireless microsensor nodes. IEEE Trans Comput 54(6):727–749
28. Gutierrez JA, Naeve M, Callaway E, Bourgeois M, Mitter V, Heile B (2001) IEEE 802.15.4: a developing standard for low-power low-cost wireless personal area networks. IEEE Netw 15(5):12–19
29. Xu Y, Heidemann J, Estrin D (2001) Geography-informed energy conservation for ad-hoc routing. In: MobiCom, Rome
30. Cerpa A, Estrin D (2004) Ascent: adaptive self-configuring sensor network topologies. IEEE Trans Mobile Comput 3(3):272–285
31. Mills KL (2007) A brief survey of self-organization in wireless sensor networks. Wirel Commun Mobile Comput 7(7):823–834
32. Pottie GJ (2001) Wireless integrated network sensors (WINS): The web gets physical. Natl Acad Eng Bridge 31(4):22–27
33. Pottie GJ, Kaiser WJ (2000) Wireless integrated network sensors. Commun ACM 43(5):51–58

Chapter 2
Sensing and Environment Concepts

Abstract This chapter provides a thorough introduction to the fundamental aspects of wireless sensor networks (WSNs), covering both theoretical concepts and practical aspects of sensing process and its role and the sensors, also, and the environment tem concept with WSN relation. The chapter also, focuses on the two sensing wireless networks relation. The WSN concepts and motivations are discussed in this chapter. It presents the sensing definitions with its devices. The relation between the wireless ad hoc network and the WSN concepts is described in this chapter.

WSN is a real time wireless network compositing a group of devices "sensors". These networks are devoted for monitoring the surround environment. WSN is wireless network can be merged in different human life activities. It can be present intelligent physical contact between the ambient mediums of the environment and the made-man medium. This physical join in general lets the specialist to monitor and control the fluctuations in the surrounding environment and to avoid the possible consequences. The tremendous advances in the WSN applications and variety of applicable fields coincides with the rapid developing in the wireless networks technology as well as the manufacturing of mechanical-microelectronics advances.

Keywords Sensing process • Environment concept • Transducers • Ecosystem balance

2.1 Introduction

The huge spreading in the WSN application is enabled by the technological advancements in wireless communications, microelectronics manufacturing and embedded computing. The wireless sensor networks is the hour subject but is not the hour discovered, it were first considered for military applications in the 1980s. The secret interest of WSN found in its capability, it is directly binding with the real world. It can transfer the real-world events and the object changing into far small room anywhere place. The Wireless sensor networks presents perfect monitoring for the surround environment and objects. This may be lead to discovering novel scientific related to the plants, the animals, or environmental variables, so, the WSN

M.A.M. El-Bendary, *Developing Security Tools of WSN and WBAN Networks Applications*, Lecture Notes in Electrical Engineering 316, DOI 10.1007/978-4-431-55069-3_2

can be considered effective scientific tool for better understanding real-world phenomena. The main and the heart of WSN is the sensing processing, it defined as gathering the information about the activity of the objects. The sensing is the first process in the WSN operation, in this chapter, it presented. If the sensor seems like the human eye, the sensing is the vision. If the sensors are considered as the human arm, the sensing is the catching or the touching. Finally, the sense of hearing, the sensor is the ears.

Figure 2.1 shows the hierarchy and contents of the wireless sensing network. The node contains the sensors and the communicating module; it is followed by the collector nodes. These nodes may be as the sensing node has another additional job. Finally, the sink or the controller, it collects and process the captured data and forward it with taking decision or not. It decides the action according to the monitored changing. The sink seems like the human brain, sensing is the process of sense, the sensors is the tools, eyes, ears, skin, and so on.

In Fig. 2.2, the WSN is not restricted by predetermined range. It can be utilized by the Aquatic biologists to monitor the sea animals are rare in the depths of oceans from small room in academic building, for example. So, the range of wireless technologies figure is compared to the environment classifications figure. The Fig. 2.2 gives more than one scenario for the sensors deployment in additional, it also, gives some of application fields of the WSN also; this figure ensures the WSN range is not restricted by the wireless technology which is utilized to implement it.

2.2 Sensing Definition

The main principle of WSN is the sensing process. The WSN and WBAN wireless networks are applied in the practically vital fields based on sensing process. So, these networks can be called "Wireless Networks Sensing Based" (WNSB). With stopping the sensing or fail to sense, it lost its establishing target. "With an interruption or failure in the sense that means that the network loses the meaning of their existence and the reason that has been lost on the basis of its creation from the beginning". Remote sensing can be defined as any process whereby information is gathered about an object, area or phenomenon without being in contact with it.

Our eyes are an excellent example of a remote sensing device. We are able to gather information about our surroundings by gauging the amount and nature of the reflectance of visible light energy from some external source (such as the sun or a light bulb) as it reflects off objects in our field of view. Contrast this with a thermometer, which must be in contact with the phenomenon it measures, and thus is not a remote sensing device as shown in Fig. 2.3.

Figure 2.4 example show the relation between the real image and the image in human eye. As example show the role of controller (sink) to translate the sensing and monitoring process to meaningful something and finding the physical meaning of the sensed material. Here, the brain play the role of sink or controller in the WSN and the eye is the remote sensor node.

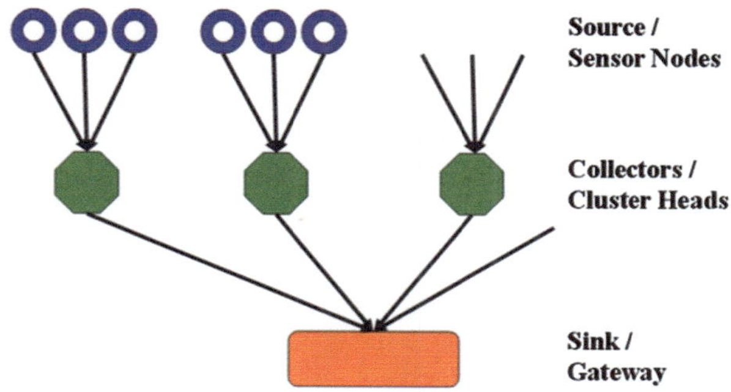

Fig. 2.1 Hierarchy of nodes and WSN contents

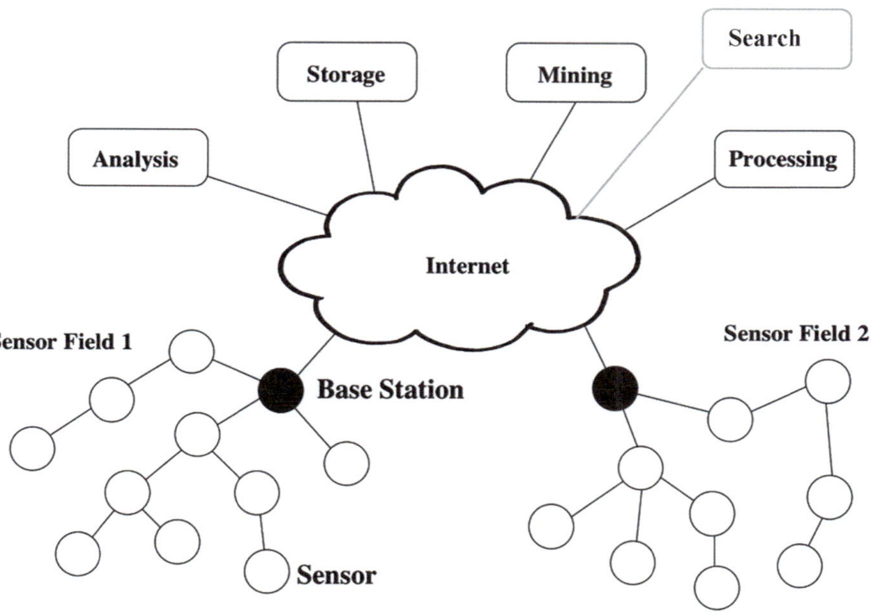

Fig. 2.2 WSN in general configuration

Sensing is the technique to gather information about physical objects or areas. Sensor (transducer): object performing a sensing task; converting one form of energy in the physical world into electrical energy. Examples of sensors from biology; the human body for real, life and closed example:

Eyes: capture optical information (light)
Ears: capture acoustic information (sound)
Nose: captures olfactory information (smell)
Skin: captures tactile information (shape, texture)

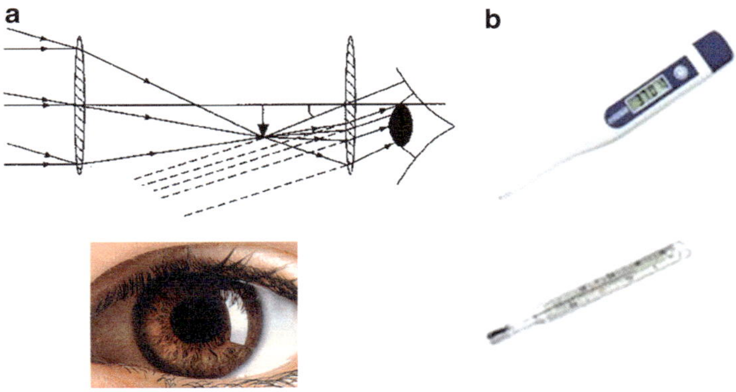

Fig. 2.3 Two examples of sensing tool with different mechanism, (**a**) Human eye and (**b**) thermometer

Fig. 2.4 Image on the retina and corresponding processed image in the brain

Sensors link the physical with the digital world by capturing and revealing real-world phenomena and converting these into a form that can be processed, stored, and acted upon. Integrated into numerous devices, machines, and environments, sensors provide a tremendous societal benefit. They can help to avoid catastrophic infrastructure failures, conserve precious natural resources, increase productivity, enhance security, and enable new applications such as context-aware systems and smart home technologies.

The phenomenal advances in technologies such as very large scale integration (VLSI), micro-electromechanical systems (MEMS), and wireless communications further contribute to the widespread use of distributed sensor systems. For example, the impressive developments in semiconductor technologies continue to produce microprocessors with increasing processing capacities, while at the same time shrinking in size. The miniaturization of computing and sensing technologies enables the development of tiny, low-power, and inexpensive sensors, actuators, and controllers. Networks of hundreds of sensor nodes are already being used to monitor large geographic areas for modeling and forecasting environmental pollution and flooding, collecting structural health information on bridges using vibration sensors, and controlling usage of water, fertilizers, and pesticides to improve crop health and quantity.

2.2.1 Definitions of Sensor and Transducer

The WSN and WBAN applications developing are related to the sensing tools. In this section, the difference between the sensor and transducer terms is presented, in general, the sensors can be considered one of the transducers types and the inversely is not true. In simple and meaningful sentence

'Sensor makes the sense *of something ⇒ to be reasonable or understandable'*

The words 'sensor' and 'transducer' are both widely used in the description of measurement systems. The former is popular in the USA whereas the latter has been used in Europe for many years. The word 'sensor' is derived from entire meaning 'to perceive' and 'transducer' is from transducer meaning 'to lead across'.

A dictionary definition of
The sensor term:
It is a device that detects a change in a physical stimulus and turns it into a signal which can be measured or recorded.

The transducer term:
It is a device that transfers power from one system to another in the same or in the different form of the transferred power.

A sensible distinction is to use 'sensor' for the sensing element itself and 'transducer' for the sensing element plus any associated circuitry. So, all the transducers contain sensors.

Figure 2.5 shows the sensing process in terms of energy conversion. The form of the output signal will often be a voltage analogous to the input signal, though sometimes it may be a wave form whose frequency is proportional to the input or a pulse train containing the information in some other form.

2.2.2 Sensor Classification

There are many ways to classify the sensors. The Sensors classification schemes can be based on the power, passive or active sensors. Also, the classification may be according to the method of sensors utilizing or based on the principle of the sensors operation. One good way to look at a sensor is to consider all of its properties, such as stimulus, specifications, physical phenomenon, and conversion mechanism, material and application field. For machine tools, sensor's conversion phenomena

Fig. 2.5 The transducer tool for sensing process

Table 2.1 Classification and examples of sensors

Sensor type	Description examples
Temperature	Thermistors, thermocouples
Pressure	Pressure gauges, barometers, ionization gauges, infrared sensors, CCD sensors
Optical	Photodiodes, photoresistors
Acoustic	Piezoelectric resonators, microphones
Mechanical	Strain gauges, tactile sensors, capacitive diaphragms, piezo-resistive cells
Motion, vibration	Accelerometers, gyroscopes, photo sensors
Flow	Anemometers, mass air flow sensors
Position	GPS, ultrasound-based sensors, infrared-based sensors, inclinometers
Electromagnetic	Hall-effect sensors, magnetometers
Chemical	pH sensors, electrochemical sensors, infrared gas sensors
Humidity	Capacitive and resistive sensors, hygrometers, MEMS-based humidity sensors
Radiation	Ionization detectors, Geiger–Mueller counters

are mainly physical phenomena such as thermoelectric, photoelectric, photomagnetic, electromagnetic, magneto-electric, thermo-elastic, thermo-magnetic, thermo-optic, photo-elastic, and so on. Stimulus is shown in Table 2.1.

Any sensor is based on a simple concept that physical property of a sensor must be altered by an external stimulus to cause that property either to produce an electric signal or to modulate an external electric signal. Quite often, the same stimulus may be measured by using quite different physical phenomena, and subsequently, by different sensors. Selection criteria depend on many factors, such as availability, cost, power consumption, environmental conditions, etc. The best choice can be done only after all variables are considered.

2.3 Environment Concepts

In this section, the relation between the WSN and the Environment has been cleared. The applicable scenario of WSN a mostly is implemented and involved in one of the environment categories as given in Fig. 2.8. The environment monitoring also, takes large part of the researchers attentions. In the following, the meaning of environment terms is presented.

a. Physical Environment

Environment is both physical and biological. It includes both living and non-living components.

"The Physical Environment is classified into three broad categories viz. (i) *Solid,* (ii) *Liquid* (iii) *Gas. Solid environment and its WSN applications, such as monitoring and predicting occurrence or the probability of occurrence of earthquakes and volcanoes or earthquakes (ground crust cracks). In order to minimize the losses that may result from them".*

b. Biological Environment

The biological of the environment consists of: *i*) Plants, (*ii*) Animals. The human and animals are greatly influenced by the climate and the weather, through Predict weather changing. Liquid and includes monitor and predict hurricanes, tidal shifts and fleecing of the seas and oceans.

Definition Environment literally means surrounding and everything that affect an organism during its lifetime is collectively known as its environment.

In another words "Environment is sum total of water, air and land interrelationships among themselves and also with the human being, other living organisms and property". It includes all the physical and biological surrounding and their interactions. Environmental studies provide an approach towards understanding the environment of our planet and the impact of human life upon the environment. Thus environment is actually global in nature, it is a multidisciplinary subject including physics, geology, geography, history, economics, physiology, biotechnology, remote sensing, geophysics, soil science and hydrology etc.

2.3.1 Scope of Environmental Science

This section is devoted to clear the Environmental science scope and find the role of WSN which is related to this science. The environmental science is a multidisciplinary science whose basic aspects have a direct relevance to every section of the society. It is the science of the physical phenomena in the environment; it studies the sources, reactions, effects and fate of physical a biological species in the air, water, soil and the effects of from the human activity upon these [1]. Its main aspects can be briefed in the following:

- Nature and natural resources.
- Biological diversity.
- Control of environmental pollution.
- Stabilization of human population and environment.

- Social issues in relation to development and environment.
- Development of non-polluting renewable energy system and providing new dimension to nation's security.

The WSN is defined as a Real-time wireless network; it can be deal with the physiological and environmental data. It provides continuous monitoring of the specific volume "service area", always real-time response also. There are many scenarios of WSN application are based on the environmental monitoring such as the agricultural applications, the sensing nodes are deployed to monitor the soil, temperature, nitrogen content, and etc. the seismic sensing application also, is considered the environmental monitoring scenario.

2.3.2 Importance of Environmental Science

Environment belongs to all the living beings and thus is, important for all. Each and every body of whatever occupation he or she may have is affected by environmental issues like global warming, depletion of ozone layer, dwindling forest, energy resources, loss of global biodiversity etc. Environment study deals with the analysis of the processes in water, air, land, soil and organisms which leads to pollute or degrade environment. It helps us for establishing standard, for safe, clean and healthy natural ecosystem. It also deals with important issues like safe and clean drinking water, hygienic living conditions and clean and fresh air, fertility of land, healthy food and development. Sustainable environmental law, business administration, environmental protection, management and environmental engineering are immerging as new career opportunities for environment protection and managements.

2.3.3 Need for Public Awareness

With the ever increasing development by modern man, large scale degradation of natural resources have been occurred, the public has to be educated about the fact that if we are degrading our environment we are actually harming ourselves. To encourage meaningful public participation and environment, it is necessary to create awareness about environment pollution and related adverse effects. The United Nations conference on Environment and Development held in Rio-de-Janeiro, followed by Earth summit on sustainable Development have high-lighted the key issues of global environmental concern and have attracted the general public towards the deteriorating environment. Any Government at its own level can't achieve the goal of environment conservation, until the public has a participatory role in it. Public participatory role is possible only when the public is awared about the ecological and environmental issues [2].

In short, if we want to manage on planet earth, we would have to make the entire population, environmentally educated. The objectives of environmental awareness should be:

(a) Improving the quality of environment.
(b) Creating an awareness among people on environmental problems and conservation.
(c) Creating such an atmosphere as people find themselves fit enough to participate in decision making process of environmental development programs.

2.4 WSN and Environment

The application of WSN for environment monitoring is essential one of WSN application. The role of WSN is not stopped at the monitoring of the surround environment or predicting the natural events happening such as the earthquake, volcanic activities and forest fire. It may be extended to the scientific activities. The utilizing of WSN within the biological life may leads to discover unknown natural phenomenon. The powerful of WSN application can be seen by its capability, it provides continuous monitoring of the specific volume. This volume includes plants, animals, land surface and etc.

In this section, this issue is discussed based on the ecosystem term concept. According to [21], "An ecosystem is the ecological unit consisting of biotic factors (living) and a biotic factor (non-living) in a specific area". For example forest, grassland, desert, aquatic etc.

Balanced Ecosystem All ecosystems, even the ultimate biosphere are open systems. There is necessarily at least an inflow and outflow of energy. Balance of ecosystem means balance of autotrophy and heterotrophy in an ecosystem, to maintain an even distribution of sustainable energy through food chain without any external interference. Any disturbance in autotrophic-heterotrophic balance leads to ecosystem imbalance, more the imbalance more externals are required to balance. Bigger the ecosystems lesser is the imbalance. Balance in fact designates to greater primary producers than consumer.

Stability of Ecosystem An ecosystem which has attained maturity is by and large is a very stable system. It is controlled by feedback mechanic who may be either positive or negative.

A simple feedback mechanism is shown below as given in Fig. 2.6:

What the relation between the WSN application and the feedback mechanism within the ecosystem?

The scenario description, here A produces B; B produces C and C produces D by the catalysis of X_1, X_2, X_3, respectively and may be a component which is needed in a little amount. When D is produced in large amount it inactivates X_1 and the reaction stops proceeding. Again D gets exhausted, simultaneously accumulation of

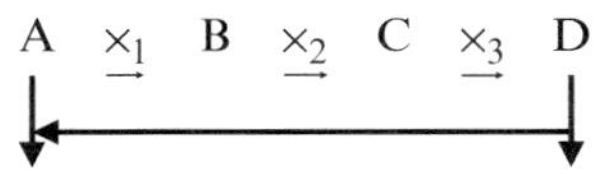

Fig. 2.6 Simple scenario of WSN utilizing for ecosystem stability

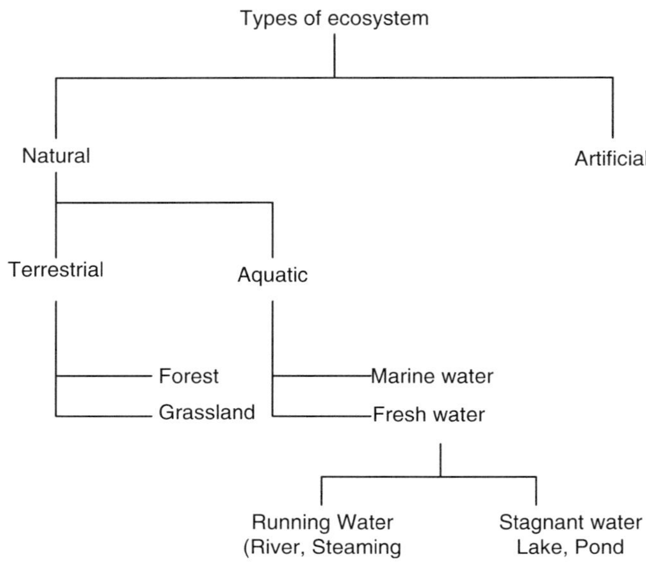

Fig. 2.7 Structure of ecosystem contents monitoring by WSN

A starts, resulting in activation of X_1. Many such feedback mechanisms occur in ecosystem to resist change. Their abilities to resist unfavorable changes have been termed as homeostatic mechanisms or, ability to return to a balanced point is homeostasis. As shown in Fig. 2.7, the Ecosystem contents can be monitored and controlled by the different implementation scenarios of the WSN.

2.4.1 Structure and Function of an Ecosystem

An ecosystem has two types of components:

1. Abiotic
2. Biotic

The WSN can be employed in many ways to contribute in the biological homeostasis, this application under the environmental monitoring WSN applicable in some fields utilizing of this wireless networks.

Fig. 2.8 WSN and the environmental monitoring

Figure 2.8 gives sample of the pollution which can be monitored and controlled to be reduced by the WSN. The Pollution from human activities and industrial which is not considered environmental factors, but the role of network sensors wireless in general and employed for environmental monitoring in both environmental factors natural or of the effects on the environment from human activities and industrial progress especially that followed certainly an increase in adverse effects on Environmental Health linked directly to human health as an individual or appointed communities, as well as the overall adverse impact on the various types of life on Earth.

After the definition of the term environment and the ecosystem meaning, as already it can be said that wireless sensors can be applied on all the contents of the physical environment of concrete. And in order to reach to reduce the dangers caused by environmental change and natural disasters, as also employs to provide superior protection for plants and crops vast widening. On the other hand the term environment also social phenomena, which may come in the future employed in detecting human behavior or societal and expected to occur the actions of the consequent damage to the environment healthy for humans, but it depends on how much progress on the discovery and invention of sensors can monitor indicators of behavior humanitarian community and then it brings.

2.5 Environment and Medical Care Concerns

In the medical sense, the environment includes the surroundings, conditions or influences that affect an organism [3]. Along these lines, last defined the environment for the International Epidemiological Association as:

> All that which is external to the human host. Can be divided into physical, biological, social, cultural, etc., any or all of which can influence health status of populations.

So based on this definition, where the role of WSN. According to this definition, the environment would include anything that is not genetic, although it could be argued that even genes are influenced by the environment in the short or long-term.

Figure 2.9 shows one way to represent the environment, from the most inclusive to the most restrictive definition.

For the purposes of environmental health, however, a more practical definition of the environment is needed, because environmental health action generally tries to change only the natural and physical environments and related behaviors (e.g. hand washing). Such interventions can rarely modify the social and cultural aspects of a community, which are usually independent of the environment (e.g. cultural pressures on lifestyle, unemployment). As a result, a more practical definition of the environment might be that given in Dif-1.

Dif-1:
A definition of "the environment" for measuring the environmental impact on the health.

The environment is all the physical, chemical and biological factors external to a person, and all the related behaviors.

- This definition excludes behavior not related to environment, as well as behavior related to the social and cultural environment, and genetics.

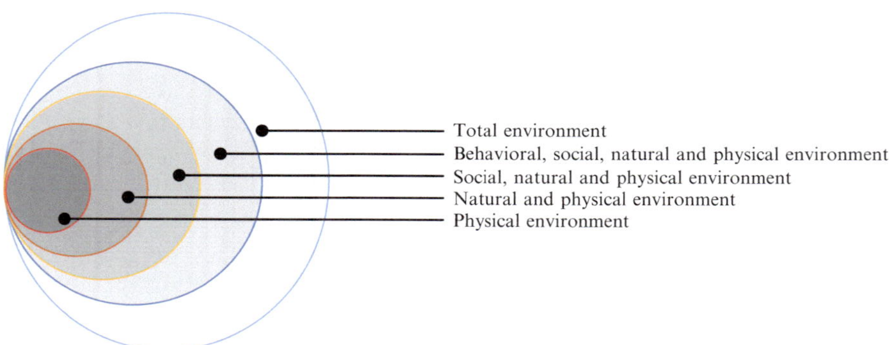

Fig. 2.9 Definition of the environment (This image adapted from Smith et al. 1999)

For our analysis, we have limited the definition of environment further, to those parts of the environment that can be modified by short-term or longer-term interventions, so as to reduce the health impact of the environment (Dif-2).

Dif-2:
In this point, there is another definition of "the environment".

The environment is all the physical, chemical and biological factors external to the human host, and all related behaviors, but excluding those natural environments that cannot reasonably be modified.

- This definition excludes behavior not related to environment, as well as behavior related to the social and cultural environment, genetics, and parts of the natural environment.

The last definition thus aims to cover those parts of the environment that can be modified by environmental management. For onchocerciasis, for example, the definition of environment would include only that part of the environment that had been affected by man-made interventions (in this case, dams), and which could be modified by further intervention. Estimates of the environmental health impact would not include disease caused by vectors living in natural environments such as rivers, if those vectors could not be controlled by reasonable environmental interventions. Similarly, deaths and injuries of soldiers during war is not included here, even though they could be considered occupational, because no intervention could possibly provide a safe working environment [3].

So, in this section of environment term definition, we tried to know the meaning of environment and how create healthy environment relation with the wide applications fields of WSN. The environmental health applications through the WSN technologies aim to preventing disease through healthy environments. For example, the WSN is presented in many researches in air pollution monitoring, Monitor the purity of drinking water, behavior of the wild animals, and etc.

2.5.1 Environmental Factors

The definition of "environment" is thus not all-inclusive in terms of the natural environment, and includes only those aspects that are modifiable. The environment contains two categories of factors, which have been included in the definition of "environment", the included environment factors which are given in the following [4].

- Pollution of air, water, or soil with chemical or biological agents;
- UV and ionizing radiation;
- Noise, electromagnetic fields;
- Occupational risks;
- Built environments, including housing, land use patterns, roads;
- Agricultural methods, irrigation schemes;
- Man-made climate change, ecosystem change;

- Behavior related to the availability of safe water and sanitation facilities, such as washing hands, and contaminating food with unsafe water or unclean hands.

The second category is the excluded environmental factors, in the following some of these factors.

- Alcohol and tobacco consumption, drug abuse;
- Diet (although it could be argued that food availability influences diet);
- The natural environments of vectors that cannot reasonably be modified (e.g. in rivers, lakes, wetlands);
- Impregnated bed nets (for this study they are considered to be non environmental interventions);
- Unemployment (provided that it is not related to environmental degradation, occupational disease, etc.);
- Natural biological agents, such as pollen in the outdoor environment;
- Person-to-person transmission that cannot reasonably be prevented through environmental interventions such as improving housing, introducing sanitary hygiene, or making improvements in the occupational environment.

These environmental factors, the WSN can be employed to modify these factors. Also, the WSN may be utilized to monitor and adjust the acceptable levels of the harmful ones.

Literary environment means the surrounding external conditions influencing development or growth of people, animal or plants; living or working conditions etc.

The familiar application of traditional sensing networks is related to the environment monitoring is the fire alarm and detection systems. The sensors of the fire alarm system monitor the changes in the specific volume. The core pieces of every fire alarm system are the same. The systems can become quite complex with multiple panels networked together over an entire campus. However the basic parts that make up the most complex system are the same as the most simple. Therefore, by learning the basics it can be built upon to assemble the parts and pieces through training for the large addressable networked systems.

2.6 WSNs and Sensors

The WSNs can be connected to the Internet so that the information can be used for future risks. Wireless sensor nodes are easy to install, remove, and replace at any location, and are expected to become smaller using Micro Electro Mechanical Systems (MEMS) technology. The WSNs will provide a ubiquitous sensing network environment in anywhere. In the hazardous condition like earthquake and forest fire, the acceleration and strain at numerous locations on each beam and column, temperature and light in each area, and images and sounds in desired regions can be obtained by the sensor nodes [5].

Additionally, a single type of sensor node such as a microphone can be used for multiple purposes. For example, a microphone can be used to detect earthquake, fires and intrusions [6]. Furthermore, a fiber optic network is not only utilized as infrastructure for information technology, but also as a "wired" sensor network. Table 2.1 shows hazards, possible applications and combination of sensors. Wireless Sensor Networks (WSN). In [7], a wireless sensor network can be classified according to application or mode of work that this requires in some specific environment. These can be classified as follows: homogeneous, self-reference, hierarchical and static. A wireless sensor network is composed of three functional levels: The level of Control, which comprises one or more control and monitoring centers, using information collected by the sensors to establish tasks that require the performance of the actuators, the level of Communications Network establishing a communication link between the field level and the level of control and the Field Level formed by a set of sensors and actuators that interact directly with the environment [8].

Due to recent technological advances, the manufacturing of small and low cost sensors became technically and economically feasible. The sensing electronics measure ambient conditions related to the environment surrounding the sensor and transform them into an electric signal. Processing such a signal reveals some properties about objects located and/or events happening in the vicinity of the sensor. A large number of these disposable sensors can be networked in many applications that require unattended operations. A Wireless Sensor Network (WSN) contains hundreds or thousands of these sensor nodes. These sensors have the ability to communicate either among each other or directly to an external base-station (BS). A greater number of sensors allows for sensing over larger geographical regions with greater accuracy [9].

2.7 Sensing Examples

The sensing process is the core of the WSN application and its main target. This process needs tool, this tool is the sensor. The sensor in WSN node is the tool of joining the network with the real world. This element likes the eye to see the surround events in the service area. It is the arm to touch the objects and translate its changing to meaningful signals [10].

Anything can be measured and its measured value is affected by number of variables, it can be indicator or sensor for anyone of these variables if other ones remain is constant. The principle operation of Thermistors or the Resistance Temperature Detector (RTD) is based on the changing of the resistance with the temperature changing, so, the resistance now is temperature sensor. The sensor is based on the resistance changing can be called "the resistive sensors" [11].

The sensors can be classified based on the method of sensing applying. As mentioned, the resistance variation with the number of variables as given in Eqs. 2.1 and 2.2. The resistive sensors are based on the electrical phenomena.

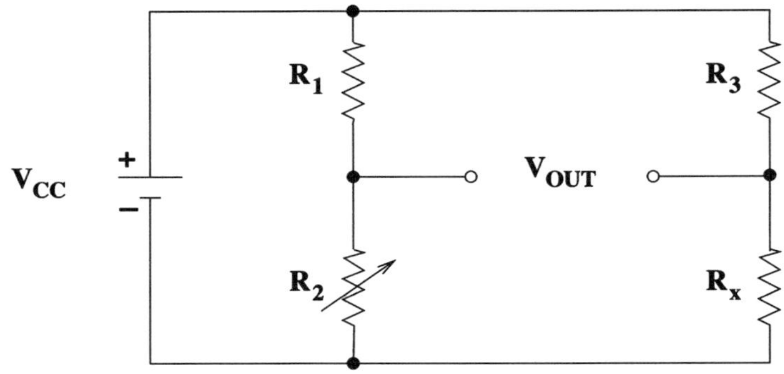

Fig. 2.10 Whitestone Bridge circuit

$$R = \frac{L \times \rho}{A} \tag{2.1}$$

The resistance in the Eq. 2.1 are affected by physical properties, where, ρ is the material conductivity, L is length of conductor, A is cross section area.

$$R = f(L,\ \rho,\ A) \tag{2.2}$$

As shown in Fig. 2.10, the Wheatstone bridge is a simple circuit that can be used to convert a physical property into an observable electric effect, the operation of the bridge is controlled by the Eqs. 2.2 and 2.3, R2/R1 and R_t/R3 value is identical, the Vout equal zero. The letter t in R_t refers to the temperature changing affects on this resistance. So, the condition of Vout = 0 is broken, the value of Vout is the indicting to the temperature value. Equation 2.4 gives the V_{out} formula.

$$\frac{R_2}{R_1} = \frac{R_T}{R_3} \tag{2.3}$$

$$V_{out} = V_{cc} \times \left(\frac{R_x}{R_3 + R_x} - \frac{R_2}{R_1 + R_2} \right) \tag{2.4}$$

The second example is the capacitor as a sensor:

A similar principle can be applied to *capacitive* sensors, which is based on the capacitance of the capacitor depends on variables as shown in Fig. 2.11. It can be used to measure motion, proximity, acceleration, pressure, electric fields, chemical compositions, and liquid depth. For example, in the parallel plate model, that is, a capacitor consisting of two parallel conductive plates separated by a dielectric with a certain permittivity ε, the capacitance is determined as:

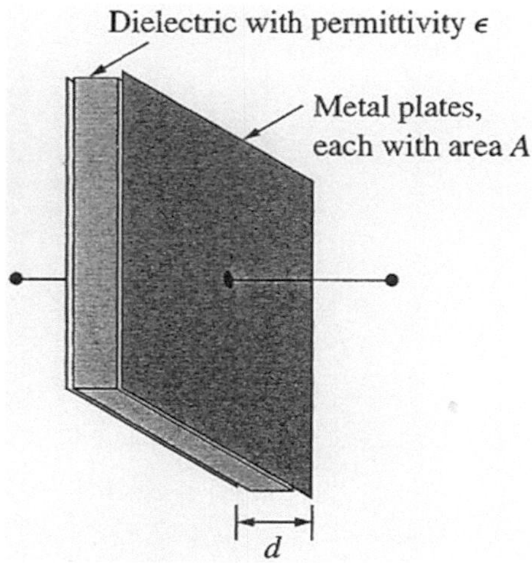

Fig. 2.11 Capacitive sensor basis, capacitor construction and variables of capacitance

$$C = \varepsilon_0 \frac{\varepsilon_r \times A}{d} \tag{2.5}$$

where A is the plate area and d is the distance between the two plates.

The variables of the capacitance equation:

C is the capacitance in farads (F),
ε_r is the relative static permittivity (dielectric constant) of the material between the plates
ε_0 is the permittivity of free space, which is equal to 8.854×10^{-12} F/m;
A is the area of each plate, in square meters and
d is the separation distance (in meters) of the two plates.

The capacitance is a function of the geometric characteristics of the capacitor - plate separation (d) and plate area (A) – and by the permittivity (ε) of the dielectric material between the plates. The permittivity of the dielectric material between the plates, it may be utilized as sensor to detect the type this material or the degree of its purity [12, 13].

$$C = f(A, \varepsilon, d) \tag{2.6}$$

Similar to the resistive model, changes in any of these parameters will change the capacitance as given in Fig. 2.12. For example, if pressure is applied to one of the two plates, the separation d can be reduced, thereby increasing the capacitance. Similarly, a change in the permittivity of the dielectric can be caused by an increase in temperature or humidity, thereby resulting in a change in capacitance. So, the capacitive sensors can be used in the humidity sensing, position sensing, proximity sensing, and the tilt sensing [14, 15].

The inductance depends on inductor's physical dimension and construction. So, the inductance can be used as sensor to indicate these physical dimensions.

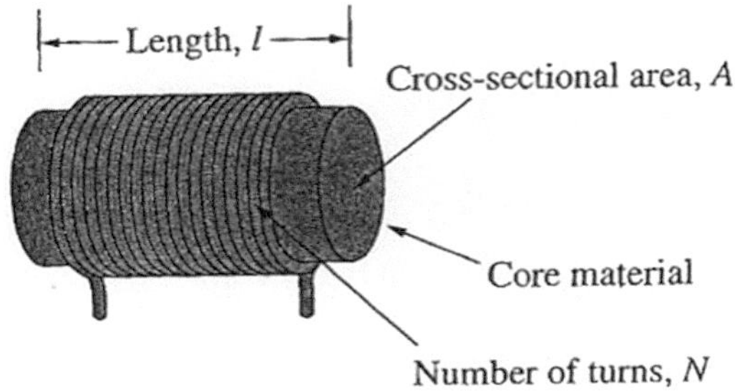

Fig. 2.12 Inductive sensor basis, inductor construction and variables of inductance

$$Ind = \frac{N^2 A\mu}{L} \tag{2.7}$$

$$Ind = f(A, \mu, L) \tag{2.8}$$

The inductance of the coil is function in number of variables, sectional area, length of coil, N is the number of turns of the coil, and the permeability of the core. *Inductive* sensors are based on the electrical principle of inductance, that is, where an electromagnetic force is induced by a fluctuating current. There are many factors determine the inductance of a coil which can be make as inductive sensor, such as the coil dimensions (Cross-sectional area, Length of coil and Number of turns = sensor dimension) the material around which it is formed. Changes in any of these parameters (e.g., caused by movements of the core within the coil) change the inductance. Inductive sensors are often used to measure proximity, position, force, pressure, temperature, and acceleration [16].

Finally, *piezoelectric* sensors use the piezoelectric effect of some materials (e.g., crystals and certain ceramics) to measure pressure, force, strain, and acceleration. When a pressure is applied to such a material, it causes a mechanical deformation and a displacement of charges, proportional to the amount of pressure. The main advantage of piezoelectric devices over other approaches is that the piezoelectric effect is not sensitive to electromagnetic fields or radiation.

Sensing is a technique used to gather information about a physical object or process, including the occurrence of events (i.e., changes in state such as a drop in temperature or pressure). An object performing such a sensing task is called a *sensor*. For example, the human body is equipped with sensors that are able to capture optical information from the environment (eyes), acoustic information such as sounds (ears), and smells (nose). These are examples of *remote sensors*, that is, they do not need to touch the monitored object to gather information as mentioned in the previous sections.

From a technical perspective, a sensor is a device that translates parameters or events in the physical world into signals that can be measured and analyzed. Another commonly used term is *transducer*, which is often used to describe a device that converts energy from one form into another. A sensor, then, is a type of transducer that converts energy in the physical world into electrical energy that can be passed to a computing system or controller [17].

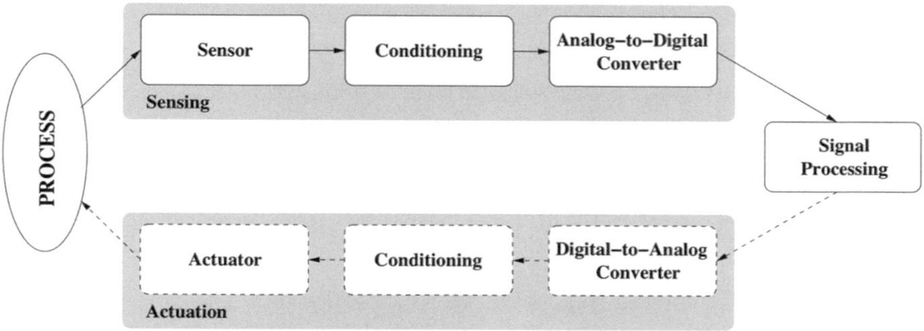

Fig. 2.13 WSN data catching and process block diagram

An example of the steps performed in a sensing (or *data acquisition*) task is shown in Fig. 2.13. Phenomena in the physical world (often referred to as *process*, *system*, or *plant*) are observed by a sensor device. The resulting electrical signals are often not ready for immediate processing; therefore they pass through a *signal conditioning* stage. Here, a variety of operations can be applied to the sensor signal to prepare it for further use. For example, signals often require amplification (or attenuation) to change the signal magnitude to better match the range of the following analog-to-digital conversion. Further, signal conditioning often applies *filters* to the signal to remove unwanted noise within certain frequency ranges (e.g., highpass filters can be used to remove 50 or 60 Hz noise picked up by surrounding power lines). After conditioning, the analog signal is transformed into a digital signal using an *analog-to-digital converter* (ADC). The signal is now available in a digital form and ready for further processing, storing, or visualization [18].

Many wireless sensor networks also include actuators which allow them to directly control the physical world. For example, an actuator can be a valve controlling the flow of hot water, a motor that opens or closes a door or window, or a pump that controls the amount of fuel injected into an engine. Such a Wireless Sensor and Actuator Network (WSAN) takes commands from the processing device (controller) and transforms these commands into input signals for the actuator, which then interacts with a physical process, thereby forming a Closed Control Loop (CCL) as shown in Fig. 2.13 [19].

The WSN scenarios of its application may act as a CCL, its works based in the interaction according to the captured events [20].

References

1. Singh YK (2006) Environmental science. New Age International Publishers, New Delhi
2. Kochtitzky C, Jackson RJ (2010) Creating a healthy environment: the impact of the built environment on public health centers for disease control and prevention. Healthy People 2010, U.S. Department of Health and Human Service

3. Davis FD (1989) Perceived usefulness, perceived ease of use, and user acceptance of information technology. MIS Q 13(3):319–340
4. Marsili D (2009) Environmental health and the multidimensional concept of development: the role of environmental epidemiology within international cooperation initiatives. Ann Ist Super Sanità 45(1):76–82
5. Serway RA, Jewett JW (2004) Physics for scientists and engineers, 9th edn 2013, Cengage learning.
6. Benenson W, Stoecker H, Harris WJ, Lutz H (2002) Handbook of physics. Springer, New York
7. Eren H, Kong WL (1999) Capacitive sensors—displacement. In: Webster JG (ed) The measurement, instrumentation, and sensors handbook. CRC Press, Boca Raton
8. Gibilisco S (2001) The illustrated dictionary of electronics. McGraw-Hill, New York
9. National Instruments Corporation (1996) Measuring temperature with RTDs – a tutorial, application note 046
10. Pallás-Areny R, Webster JG (2001) Reactance variation and electromagnetic sensors. In: Sensors and signal conditioning. Wiley, New York, pp 207–273
11. Ripka P, Tipek A (2007) Level position and distance. In: Modern sensors handbook. ISTE USA, Newport Beach, pp 305–348
12. Jerančе N, Bednar N, Stojanović G (2013) An ink-jet printed Eddy current position sensor. Sensors 13:5205–5219. doi:10.3390/s130405205
13. Samatham R, Kim KJ, Dogruer D, Choi HR, Konyo M (2007) Active polymers: an overview. In: Kim KJ, Tadokoro S (eds) Electroactive polymers for robotic applications: artificial muscles and sensors. Springer, London, p 18
14. Gründler P (2007) Conductivity sensors and capacitive sensors. In: Chemical sensors. Springer, Berlin
15. Baxter LK (1997) Capacitive sensors—design and applications. In: Herrick RJ (ed) Wiley-IEEE Press
16. Dadafshar M (2005) Inductor current-sensing boosts regulator efficiency. Power Electronics Technology. www.powerelectronics.com
17. Pallás-Areny R, Webster JG (2001) Sensors and signal conditioning. Wiley, New York
18. Kilian CT (2000) Sensors. In: Modern control technology: components and systems. Delmar Thomson Learning, Novato, pp 220–294
19. Tansley AG (1935) The use and abuse of vegetational concepts and terms. Ecology 16:284–307
20. Kuttruff H (1991) Ultrasonics—fundamentals and applications. Elsevier Applied Science, London

Chapter 3
Wireless Networks

Abstract The developing and spreading wireless sensing applications based on the tremendous developing of the different wireless networks and technologies. This chapter gives a brief study for the different wireless networks types based on the range. Also, it presents with more details the most common acceptable and applicable wireless networks within the WSN applications that is the short range wireless networks. Also, it describes the Bluetooth and ZigBee networks in detail. Many of WPANs technologies are used to implement the WSN networks. This chapter focused on the common used WPANs technologies. The adaptive packets technique is used to enhance the performance and improve the throughput over the wireless link. This point is described based on the Bluetooth technology packets.

Keywords WPANs • Bluetooth • ZigBee • Adaptive packet

3.1 Introduction

The WPANs are involved in several essential applications such as the military services and industrial control. In general, the performance of wireless networks is degraded by factors like signal attenuation, reflection, multipath effect, and man-made interference. So, the QoS of this type of important wireless networks is a vital target. On the other hand, several applications may need multimedia signal transmission. In addition, the digital communication techniques have been developed so fast that people are more and more eager for multimedia communication. The most popular WPANs technologies are wireless Ad-hoc networks. The Ad-hoc is a Latin phrase that means auto-configuration and flexibility. Figure 3.1 gives the infrastructure of the traditional Ad-hoc networks.

In this chapter, Sect. 3.2 gives an overview of low-rate wireless networks. In Sect. 3.3, we present the Bluetooth network as a widely used WPAN network, and its different packet formats. In Sect. 3.4, a discussion of the adaptive packets of WPAN technology and the effect of size variability of Bluetooth packets on the Bluetooth network performance is introduced.

M.A.M. El-Bendary, *Developing Security Tools of WSN and WBAN Networks Applications*, Lecture Notes in Electrical Engineering 316, DOI 10.1007/978-4-431-55069-3_3

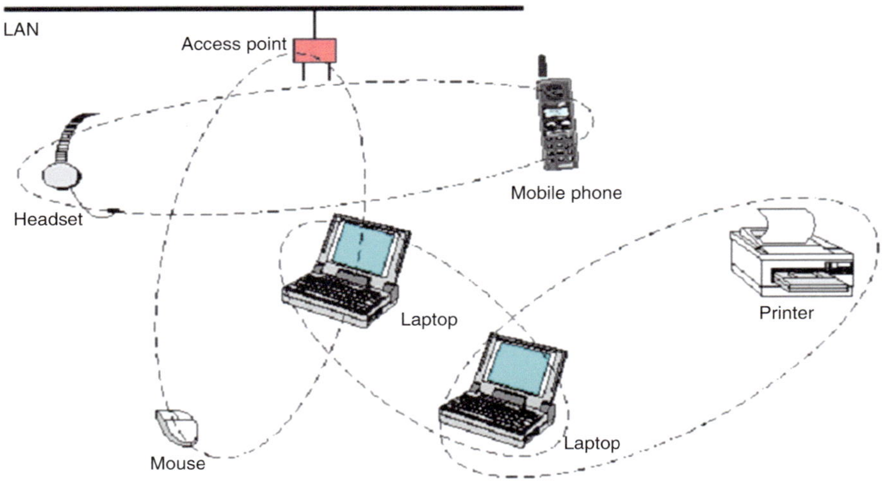

Fig. 3.1 WPAN network example, a scatternet with four piconets

3.2 Low-Rate Wireless Networks

We will focus in this book on two low-rate wireless technologies; Bluetooth and ZigBee.

3.2.1 Bluetooth Technology

Bluetooth, promoted by the Bluetooth Special Industrial Group (SIG), provides a comparably low-cost solution for wireless communication among portable or handheld devices at a maximum data rate of 1 Mbps within up to 100 m [1]. It operates in the 3.4 GHz ISM band with as low as 0 dBm transmission power, and employs Frequency Hopping Spread-Spectrum (FHSS) techniques to overcome interference and multi-path fading in the wireless channel. Meanwhile, Bluetooth adopts FEC and Automatic Repeat reQuest (ARQ) to improve reliability by reducing errors in data transmission. A standard Bluetooth network called piconet consists of a group of Bluetooth enabled devices sharing the same communication channel. Figure 3.1 shows the piconet and scatternet structure. In the piconet, only one device serves as the master unit that actively synchronizes other devices, whereas the rest of devices serve as slave units that passively establish the communication with the master unit. Technically, a master unit is able to interact with up to 7 slave units at the same time and synchronizes more than 200 slave units without communication. The second type of Bluetooth networks is the scatternet, which is formed by more than one piconet [2].

In order to avoid interference from the same frequency band, Bluetooth adopts the Time Division Multiple Access (TDMA) technique to allow different time slots [3].

3.2.2 ZigBee/IEEE 803.15.4 Technology

ZigBee/IEEE 803.15.4 [4] is a bidirectional wireless technology featured with short range, low cost, low power consumption, and low data rate, which makes it more suitable for applications associated with monitoring and remote control that are integrated with functional sensors and actuators. Normally, ZigBee works in the license-free 3.4 GHz ISM band with a data rate of up to 250 Kbps and the transmission distance ranges from 10 to 75 m, depending on the power output and environmental dynamics [5]. ZigBee entirely adopts the IEEE 803.15.4 physical (PHY) layer and Medium Access Control (MAC) layer as the underlying layers to support reliable data transmission in a harsh environment with noise and signal disturbances as illustrated in Fig. 3.2.

At the PHY layer, ZigBee uses Direct-Sequence Spread Spectrum (DSSS) with two different Phase-Shift Keying (PSK) modulations to minimize interference. At the MAC layer, ZigBee adopts the Carrier Sense Multiple Access/Carrier Access (CSMA/CA) mechanism to improve network throughput and minimize transmission delay [5]. The main advantages of the ZigBee technology are summarized as follows [6]:

1. Low power consumption.
2. Low data rate (250 Kbps).
3. Short distance.
4. Low cost, resulting from the low-rate and the simplicity of the protocol stack.

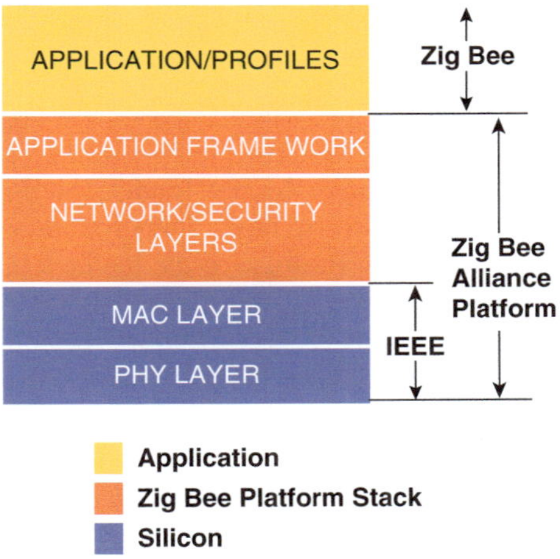

Fig. 3.2 The architecture of ZigBee layers

Table 3.1 Technical comparison between Bluetooth and ZigBee

Comparing wireless technologies		
Features	IEEE 803.15.3	IEEE 803.15.4
Battery life	Days	Years
Complexity	Complex	Simple
Nodes/master	7	65,540
Range	Up to 100 m	75 m
Base data rate	1 Mbps	250 Kbps
Effective throughput	700 Kbps	100 Kbps
Applications	File transfer	Monitoring and control

5. Possibility of up to 65,535 devices per network.
6. Robustness and reliable data transfer.
7. Flexibility in networking with multiple topologies.
8. Data integrity verification and authentication by adopting the 128-bits AES encryption algorithm at the MAC layer.

In comparison with ZigBee, the main disadvantages of Bluetooth technology are the price of purchase, the number of network nodes, the limited distance along with the corresponding power consumption in terms of the coverage of home control networks as Table 3.1 shows:

Given the characteristics of ZigBee, the authors in [7] built an experimental environment for smart home solutions including energy consumption management, entrance guards, indoor security, home automation and real-time message delivery. What is important here is that the home appliances are specially equipped with a microcontroller and sensors in such a way to exchange appliance status information and instructions for control between appliances and ZigBee nodes. Another similar but simplified solution with sensors and actuators, specially intended for environment dynamics, was built in [8].

3.3 Bluetooth Networks Architecture

At the beginning of the Bluetooth technology, it was considered as a natural WLAN extension. With the rapid spread of this technology, it has become involved in different fields. Because of the features of the Bluetooth technology as an end-to-end wireless solution, many researchers have suggested its use for remote control in a smart home. This proposition was established by the authors in [9]. Also, it can be used in cooperation with WLANs or wired LANs as shown in Fig. 3.3.

One of the most attractive features of Bluetooth technology is the very low power consumption that permits its integration in portable, battery driven electronic devices, such as mobile phone, and mouse. As a matter of fact, Bluetooth standard defines four operational modes, namely Active, Hold, Sniff, and Parked.

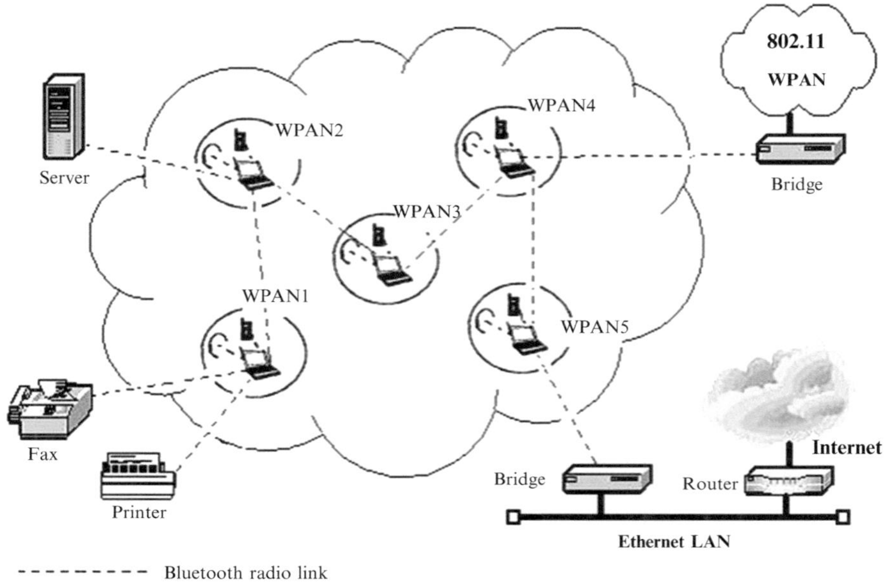

Fig. 3.3 Bluetooth in cooperation with other networks

These modes correspond to different degrees of activity and, in turn, different levels of power consumption. Besides these high-level mechanisms, energy-saving is also pursued at a microscopic level, by means of a suitable packet reception mechanism that permits a device to switch off the receiver circuitry as soon as it realizes that the incoming signal cannot be correctly decoded or it is addressed to another device. In this way, a unit that is not addressed by any valid packet is active for less than 10 % of the time [10].

Despite this very attractive low-power feature, some implementation and compatibility problems have slowed down the penetration of the Bluetooth technology in the market, until recently. Most of such problems are now solved and Bluetooth is undertaking the expected success, being integrated in hundreds of portable electronic devices. However, the first generation of Bluetooth products, compliant with the Bluetooth v1.0 specifications [11], was characterized by a low transmission rate (1 Mbps) and a rather long connection set-up time (order of seconds), which have restricted the use of Bluetooth mainly to cable- replacement applications. These limitations have been partially removed by the enhancements included in the Bluetooth v3.0 + EDR specifications [12]. The new version of the standard, in fact, includes an EDR mode for higher transmission rates (up to 3 Mbps) together with other improvements aimed at speeding up the node discovery and connection setup procedures and limiting the interference with other devices operating in the same frequency band [13].

In February 2007, moreover, the Bluetooth SIG published the last update of the standard, Bluetooth v3.1 + EDR [14], which contains further improvements to the link establishment procedure. With these upgrades, Bluetooth becomes ready to break through the borders of cable-replacement products and enter the wide arena of high-speed radio technologies and pervasive networks. This enlarged and challenging scenario includes different types of applications, such as opportunistic data exchange, bulk data transfer, distributed and cooperative computing and storage and so on [15].

The success of Bluetooth in these competitive areas, however, depends on the actual performance, in terms of throughput, delay and energy efficiency, that the technology can provide in realistic propagation environments. This topic has been partially addressed by previous work. In [16], the authors investigate some techniques to improve Bluetooth EDR data throughput by using FEC and interleaving schemes. They consider a binary numeric channel model where bit errors occur independently or according with a Gilbert model, and they determine the throughput achieved by each baseband frame format for a given bit error probability, using different error correction mechanisms. However, the study does not present any delay and energy efficiency analysis.

The point-to-point Bluetooth throughput achieved by 1 Mbps frame formats is derived in [17] for different channel conditions. A mathematical approach to performance analysis of Bluetooth piconet can be found, for instance, in [18]. The aim of such works, however, is to model the general performance trend of the system, in order to permit a comparative analysis of different polling and retransmission strategies, rather than providing accurate results for throughput and energy efficiency. In fact, the models are often based on simplifying hypotheses, which are required to make the problem tractable, and neglect some details of the Bluetooth technology that impact on the system throughput and energy efficiency. Segmentation-and-Reassembly (SAR) policies are investigated in [19], where throughput is still considered as the only performance metric.

3.3.1 Bluetooth v3.0 + EDR

This section shortly overviews the features of the Bluetooth v3.0 + EDR standard that are of interest for our analysis. An introductory description of the main features of Bluetooth technology can be found in [20].

3.3.1.1 Physical Layer (Basic and Enhanced Rates)

The Bluetooth v3.0 + EDR specifications encompass three modulation schemes, which correspond to a basic data rate of $R_1 = 1$ Mbps (BR), and two enhanced data rate modes of $R_2 = 2$ Mbps (2EDR) and $R_3 = 3$ Mbps (3EDR), respectively. The BR makes use of a binary Gaussian-shape Frequency Shift Keying (GFSK) scheme, while 2EDR and 3EDR are obtained by using Differential-encoded Phase Shift Keying (DPSK) modulations, with a constellation of four symbols (π/4-DQPSK) and eight symbols (8DPSK), respectively [21]. In all the cases, the symbol period remains equal to $T_s = 1$ μs, so that the frequency band of the radio signal is not significantly modified by the introduction of the EDR schemes. For a given Signal-to-Noise Ratio (SNR), defined as the ratio between the average energy per symbol E_s and the noise energy N_0, the BER of GFSK and π/4-DQPSK modulations is very similar, so that the last scheme is always preferable, giving a transmission rate that is twice the basic one.

This performance gain is paid in terms of transceiver complexity. For instance, the basic-rate GFSK scheme, being a constant-envelope modulation, permits to have the amplifier working in proximity of the saturation point, where it is most efficient. Conversely, DPSK modulation schemes have a peak to average ratio of about 3.3 dB, which requires moving the working point of the amplifier below the saturation point, in order to avoid clipping effects. Therefore, to maintain the same output power, more efficient amplifiers have to be used [22]. In the light of this consideration, the rest of this chapter will be focused on EDR schemes.

3.3.1.2 ACL Baseband Frame Formats

The Bluetooth standard encompasses two types of links; Synchronous Connection Oriented (SCO) and Asynchronous Connection-Less (ACL). SCO links are aimed at the transport of delay-sensitive traffic (mainly voice) and make use of a periodical time-reservation scheme [23]. ACL links are intended for the transport of asynchronous data traffic, as file transfer and web browsing. We will focus on ACL links only. Bluetooth v3.0 + EDR adds six ACL frame formats to the basic rate formats introduced in the first version of the standard. Each ACL data frame begins with an Access Code (AC) field that is used for synchronization; DC offset compensation and piconet identification. AC is followed by the frame header (HD) field, which contains link control information, including frame type, destination address, sequence number, and acknowledgment flag [24].

Furthermore, the HEAD field contains a checksum word (HEC), which is used to verify the integrity of the field after decoding. For backward compatibility, AC and HEAD fields are always transmitted at the basic rate [25]. In EDR frames, the HEAD is followed by a guard time of approximately 5 μs, which is used to switch the transceiver circuitry to the appropriate DPSK scheme. The guard time is followed by a synchronization field (SYNC) of 10 DPSK-modulated symbols that is used for signal acquisition at the receiver. The SYNC is followed by a variable length Payload (PL) field, which still includes 2-byte header and 2-byte CRC. The last field of the frame is a trailer field of only two symbols. SYNC, PL and Trailer fields are transmitted by using the selected EDR modulation scheme. The time occupancy of EDR frames is limited to 1, 3 or 5 consecutive slots. The different EDR frame formats are denoted by jDH_x, where DH refers to High Data rate, $j = 2$, 3 is the transmission rate (in Mbps), while $x = 1, 3, 5$ is the slot occupancy. Besides these data frames, Bluetooth specifications define two short control frames, named POLL and NULL, which contain AC and HEAD fields only with no PL.

Table 3.2 Bluetooth SCO packet types

Packet types	HV_1	HV_2	HV_3	DV
Packet header (bytes)	None	None	None	1D
Payload length (bytes)	10	20	30	20
Channel utilization (%)	100	50	33	100
FEC code	1/3	2/3	None	2/3D

Table 3.3 Bluetooth ACL packet types

Packet types	DH_1	DM_1	DH_3	DM_3	DH_5	DM_5
Slot time	1	1	3	3	5	5
Packet header (bytes)	1	1	2	2	2	2
Payload length (bytes)	0–27	0–17	0–183	0–121	0–224	0–339
FEC code	None	2/3	None	2/3	None	2/3

Table 3.4 Bluetooth ACL double rate EDR packet types

Packet types	$2DH_1$	$2DM_1$	$2DH_3$	$2DM_3$	$2DH_5$	$2DM_5$
Slot time	1	1	3	3	5	5
Packet header (bytes)	1	1	2	2	2	2
Payload length (bytes)	0–54	0–34	0–367	0–244	0–679	0–450
FEC code	None	2/3	None	2/3	None	2/3

Table 3.5 Bluetooth ACL triple rate EDR packet types

Packet types	$3DH_1$	$3DM_1$	$3DH_3$	$3DM_3$	$3DH_5$	$3DM_5$
Slot time	1	1	3	3	5	5
Packet header (bytes)	1	1	2	2	2	2
Payload length (bytes)	0–83	0–52	0–552	0–367	0–1,021	0–676
FEC code	None	2/3	None	2/3	None	2/3

The characteristics of the different frame formats are summarized in the Tables 3.2, 3.3, 3.4, and 3.5.

3.3.1.3 Baseband

The basic Bluetooth network configuration is the so-called piconet, a cluster of no more than eight devices sharing a common frequency-hopping radio channel. When the piconet is established, one unit gets the master role, while the others get the slave role. The master is in charge to manage the medium access by means of a polling scheme, the master cyclically polls the slave by sending either useful data frames or POLL frames [26]. The slave addressed by the master polling is required to immediately reply by transmitting a data frame or a NULL frame.

Bluetooth provides a reliable data connection by using an ARQ mechanism at the baseband layer. Each data frame is transmitted and retransmitted until the source node gets a positive acknowledgment (ACK) from the destination. The ACK is carried in the HEAD field of the baseband frame (piggy-backing), so that its reception probability is independent of the frame format. Since negative ACK is assumed by default, master retransmissions are also triggered by ACK losses. In particular, the loss of frames carrying positive ACKs will trigger the retransmission of frames that were already successfully delivered to the slave. These frames, which are called duplicate packets (DUPCKs), can be easily recognized by the slave since they all carry the same sequence number in the HEAD field. Slaves disregard the PL field of DUPCKs, always returning a positive ACK. Notice that, slave transmissions are allowed only upon receiving a valid master (polling) frame that, in turn, will also carry the positive or negative acknowledgment for the previous slave-to-master transmission. Therefore, slaves retransmissions occur only when solicited by an explicit not acknowledgment sent by the master [14].

3.3.2 *Micro-level Energy Saving Mechanisms*

Energy-saving was a key feature in the design of the Bluetooth technology. According to this principle, a unit stops receiving and enters a low-power doze mode as soon as it determines that a field in the incoming frame is affected by unrecoverable errors or the frame is addressed to another unit or the signal strength is too low to guarantee good reception. More specifically, at the beginning of each receive slot, the Bluetooth unit scans the received radio signal looking for a valid AC field. If the AC is not recognized within a proper time window, reception stops and the unit enters a low-power doze mode until the beginning of the following receive slot. Conversely, after the recognition of the AC field, the receiver processes the HEAD field and checks the validity of the HEC word. If the check fails, the device enters doze-mode, otherwise the HEAD field is inspected to determine the frame format and the destination address. Slaves not addressed by the master transmission may enter doze-mode till the end of the frame. The slave addressed by the master, instead, checks the sequence number contained in the HEAD field of the incoming frame to verify whether it is a DUPCK or not. In the first case, the remaining of the frame is not decoded and the slave enters doze mode till the end of the transmission, after which it piggy-backs a positive acknowledgment to the master. When the incoming frame is not a DUPCK, instead, the slave decodes the entire frame and, then, it piggy backs a positive or negative ACK according to the outcome of the CRC of the PL field [27].

3.4 Adaptive Packets of WPAN Technology

The key idea for adaptive packet selection and scheduling methods is to adapt the transmission according to channel conditions. For instance, if the channel is dominated by interference from 803.11b network, packet loss will be mainly due to collisions between IEEE 803.15.1 and 803.11 systems, instead of bit errors resulting from noise. Packet types that do not include FEC protection could provide better throughput if combined with intelligent packet scheduling. The foundation for the effectiveness of these types of methods is to be able to figure out the current channel conditions accurately and timely [28].

In this book, we introduce several methods that enhance the performance of the IEEE 803.15.1 and 803.11 networks through the use of adaptive packet selection and scheduling for the Bluetooth devices. These methods do not require the collaboration between the 803.11 devices and the Bluetooth devices. Therefore, they belong to the general category of non-collaborative coexistence mechanisms.

The ACL link, in addition to the use of different FEC protections, adds the choice of multi-slot packets. The following tables summarize the packets for ACL link. Again the different ACL packets allow the applications to make trade-offs among different considerations of traffic flow, channel conditions of the current hop, duty cycles and interference generated to neighboring network [29].

As cleared by these Tables 3.2, 3.3, 3.4, and 3.5, the Bluetooth technology is adaptive packets technology; it handles the utilized packet for transmission process based on the channel condition. The Bluetooth specifies a variety of packet types with different combinations of payload lengths, slots occupied, FEC codes, and ARQ options. The motivation there is to provide the necessary flexibility for the implementers and applications so that the packets can be chosen optimized for the traffic and channels presented. In this book, we describe a mechanism to take advantage of these different packet types for improving network capacity for coexistence scenarios. Bluetooth provides four types of packets that can be sent over an SCO link; HV_1, HV_2, HV_3, and DV packets [30].

The different packets differ mostly in the FEC code used and the amount of channel occupied by the SCO link. Choice of different packet types provides intriguing trade-offs of error protection at the bit level and the amount of interference generated or the bandwidth available for other links [31].

3.4.1 Methods of Adaptive Packet Selection

As mentioned in previous section, Bluetooth provides four types of packets that can be sent over an SCO link: HV_1, HV_2, HV_3, and DV packet. The different packets differ mostly in FEC code used and the amount of channel occupied by the SCO link. Choice of different packet types provides intriguing tradeoffs of error protection at the bit level and the amount of interference generated (or the bandwidth available for other links) [32].

The basic idea here is to adaptively select packet types, given either an ACL or SCO link, so that maximal total network capacity is achieved. This implies not only optimizing throughput for the Bluetooth piconet but also reducing interference to the coexisting 803.11b network, which will increase the throughput of the 803.11b network. For SCO links, when the network performance is range limited, that is, the stations are separated by a distance such that only small noise margin is maintained. In this case, random bit errors are the dominant problem for dropping packets. By choosing a packet type that uses more error protection will increase the performance of the SCO link. Therefore, for range limited applications, HV_1 packet is preferred over HV_2 packet, and HV_2 packet is preferred over HV_3 packet. By monitoring the RSSI and SNR of the Bluetooth radio, the Bluetooth can determine if the choice of more error protection is beneficial. For SCO links in the coexistence scenarios, usually the dominant reason for packet drop is not due to noise or range, but rather is due to the strong interference produced by the collocated network such as 803.11b network [33].

In this case, by increasing FEC protection will cause Bluetooth device to generate more packets (HV_1 packets occupy the channel three times more often than HV_3 packets), and thus a lot more interference to the 803.11b network. As shown by the simulation results in the following figure, the total network throughput is severely degraded. The figure demonstrated the performance of the 803.11b network before and after the initiation of an HV1 SCO link by the Bluetooth piconet. The 803.11b throughput dropped from 5.8 Mbps to around 1 Mbps. Therefore, in interference-limited scenarios (as in Bluetooth and 803.11b coexistence scenarios), we should prefer HV_3 packet over HV_2 packet, and HV_2 packet over HV_1 packet [34].

The reason of different voice transmitting packets format is to allow the SCO link the flexibility of choosing the hops that are out-of-band with the collocating 803.11b network spectrum for transmission. The duty cycle or channel utilization of the SCO link does not change. For HV_3 packets, we will send one HV_3 packet every six slots, which uses 33 % of the total available bandwidth. The only change proposed is to allow the piconet master the flexibility of choosing when to initiate the transmission.

For similar reasons, the same guidelines apply to the selection of ACL packets. When the Bluetooth network performance is range-limited, we should use ACL packets with FEC protections, which include DM_1, DM_3 and DM_5. On the other hand, when the system is interference limited, we should try to reduce the number of bits transmitted by choosing more bandwidth efficient packet format such as DH_1, DH_3 or DH_5 [35].

The previous section discusses the reasons of the variety of audio signals and data transmission packets. This packets variety provides the following, maximum channel capacity according to the channel of communications conditions. Also, there is no a protocol manages the co-operation between different wireless network which are working in the same frequency band, such as WLAN IEEE 803.11b,

ZigBee IEEE 803.15.4, and Bluetooth IEEE 803.15.1. It is can be said that the changing of packets length capabilities may decrease the probability of collision, hence decreasing the amount of dropped and lost data.

In addition, the variety of packets format may lead to reduce the computational complexity of the system through using no FEC packets in case of good condition channel.

References

1. Mohamed MAM, Abou El-Azm A, El-Fishwy N, El-Tokhy MAR, Abd El-Samie FE (2008) Optimization of Bluetooth packet format for efficient performance. Prog Electromagn Res M 1:101–110
2. Zanella A (2009) A mathematical framework for the performance analysis of bluetooth with enhanced data rate. IEEE Trans Commun 57(8):2463–2473
3. Mohamed MAM, Abou El-Azm AE, El-Fishawy NA, El-Tokhy MAR, Shawky F, Abd El-Samie FE (2009) Bluetooth performance improvement using convolutional codes. J Autom Syst Eng 3(1)
4. Tan T, Huang YF, Chiu H, Chang C (2009) Enhancing Bluetooth data throughput using a mode/packet switching scheme based on ARQN statistics. In: International proceedings of the eighth international conference on machine learning and cybernetics, Baoding, 12–15 July 2009
5. Chen JJ, Sun T (2006) Improving Bluetooth throughput using FEC and interleaving. In: International conference on mobile ad hoc sensor networks, Hong Kong, pp 726–736
6. Valenti MC (2002) On the throughput of Bluetooth data transmissions. In: IEEE wireless communications and networking conference. IEEE Communications Society pp 119–123
7. Hamdi KA (2002) Exact probability of error of BPSK communication links subjected to asynchronous interference in rayleigh fading environment. IEEE Trans Commun 50 (10):1577–1579
8. Conti A, Dardari D, Paolini G, Andrisano O (2003) Bluetooth and IEEE 802.11b coexistence: analytical performance evaluation in fading channels. IEEE Trans Sel Areas Commun 21(2):259–269
9. IEEE 802.11, the working group setting the standards for wireless LANS. http://grouper.ieee.org/groups/802/11
10. El-Bendary MAM, Abu El-Azm AE, El-Fishawy NA, Hagagg A, Shawky F (2009) Throughput improvement over bluetooth system through adaptive packets. In: The 2009 international conference on computer engineering & systems (ICCES'09), Triumph Hotel, Cairo, 14–16 Dec 2009
11. Miller LE, Lee JS (1998) BER expressions for differentially detected π/4 DQPSK modulation. IEEE Trans Commun 46(1):71–81
12. Mohamed MAM, Abou El-Azm A, El-Fishwy N, El-Tokhy MAR, Abd El-Samie FE, Shawki F (2008) Bluetooth performance improvement with existing convolutional codes over AWGN channel. In: Proceedings of the 2nd international conference on electrical engineering design and technologies (ICEEDT'08), Transaction No. 537, Hammamet, 8–10 Nov 2008
13. Golmie N, Van Dck RE, Soltanian A (2001) Interference of Bluetooth and IEEE 802.11: simulation modeling and performance evaluation. In: Proceedings ACM international workshop on modeling, analysis, and simulation of wireless and mobile systems, Italy
14. Miorandi D, Zanella A (2004) Achievable rate regions for Bluetooth piconets in fading channels. In: Proceedings of the IEEE VTC (Spring), Milan

15. Chui TY, Thaler F, Scanlon WG (2002) A novel channel modeling technique for performance analysis of Bluetooth baseband packets. In: Proceedings of the IEEE ICC conference, New York
16. Haartsen JC, Zürbes S (1999) Bluetooth voice and data performance in 802.11 DS WLAN environment. Ericsson report
17. Lentmaier M, Zigangirov KSh (1999) On generalized low-density parity-check codes based on hamming component codes. IEEE Commun Lett 3(8):248–250
18. Kaiser M, Fong W, Sikora M (2009) A comparison of decoding latency for block and convolutional codes. In: Proceeding, ISCTA'09, Ambleside
19. Hagenauer J, Papke L (1996) Iterative decoding of binary block and convolutional codes. IEEE Trans Inf Theory 42(2)
20. Specification of Bluetooth system, ver. 1.1, 22 Feb 2001
21. Specification volume 2: Core system package, Bluetooth specification version 2.0 + EDR ed. The Bluetooth Special Interest Group, 4 Nov 2004
22. Specification volume 2: Core system package, bluetooth specification version 2.1 + EDR ed. The Bluetooth Special Interest Group, 26 July 2007
23. Pasolini G, Chiani M, Verdone R (2002) Proposal of a MAC strategy for a Bluetooth based WLAN and performance evaluation in realistic channel conditions. In: Proceedings of PIMRC'2001, San Diego, Sept 2001. International Journal of Wireless Information Networks
24. Valenti M, Robert M, Reed J (2002) On the throughput of Bluetooth data transmission. In: Proceedings of the IEEE WCNC, Orlando
25. Miorandi D, Zanella A, Pierobon G (2004) Performance evaluation of Bluetooth polling schemes: an analytical approach. Mobile Netw Appl 9(1):63–72
26. Miorandi D, Caimi C, Zanella A (2003) Performance characterization of a Bluetooth piconet with multi–slot packets. In: Proceedings of the WiOpt03, Sophia Antipolis
27. Zussman AYU, Segall G (2007) On the analysis of the Bluetooth time division duplex mechanism. Wirel Commun IEEE Trans 6(6):2149–2161
28. Ling-Jyh C, Tony S, Chen Y-C (2006) Improving Bluetooth EDR data throughput using FEC and interleaving. In: The second international conference on mobile ad hoc and sensor networks (MSN'06). Hong Kong, China pp 725–736
29. Misic V, Misic J (2003) Modeling Bluetooth piconet performance. Commun Lett IEEE 7 (1):18–20
30. Misic J, Misic VB (2002) On Bluetooth piconet traffic performance. In: Personal, indoor and mobile radio communications, 2002. The 13th IEEE international symposium on, 15–18 Sept 2002, vol 1, pp 501–505
31. Kalia M, Bansal D, Shorey R (1999) MAC scheduling and SAR policies for Bluetooth: a master driven TDD pico-cellular wireless system. In: Proceedings of the IEEE international workshop on mobile multimedia communications (MoMuC'99). San Diego, CA Nov 1999
32. Jang K, Lee T-J, Kang H, Park J (2001) Efficient power management policy in Bluetooth. IEICE Trans Commun E84-B(8):2186–2192
33. Haartsen J (2000) Bluetooth towards ubiquitous wireless connectivity. Revue HF, Soc Belge Ing Telecommun Electron 8–16
34. Roh JS (2006) Performance analysis and evaluation of Bluetooth networks in wireless channel environment. In: ICSNC'06: proceedings of the international conference on systems and networks communication. IEEE Computer Society, Washington, DC, pp 61–65
35. Miller LE, Lee JS (1998) BER expressions for differentially detected Л/4 DQPSK modulation. IEEE Trans Commun 46(1):71–81

Chapter 4
Throughput Study of Low-Rate Communications

Abstract As known the WSN application have many of challenges, the most important one of them is the power. The challenge of minimizing transmit power is deserved to be the most critical parameter for a low-power WSN, whose battery lifetime is dependent on power consumption. Therefore all transmit power and energy calculations use the minimum required transmit power and energy. In a low-power WSN scenario, transmitting with as much power as possible, up to regulatory limits, is not desirable. Rather, transmitting with as little power as possible, so as to extend sensor battery life, while maintaining a minimum required SNR, is our goal. Similar to a deep space satellite scenario, the low-power WSN is far more power-constrained than bandwidth-constrained. In order to achieve power efficiency, we are willing to sacrifice spectral efficiency.

Keywords Area throughput • Throughput • WSN topology • Adaptive packets

4.1 Introduction

The chapter studies a vital issue in wireless communications, which is the throughput of the transmitted data over wireless networks. The chapter presents different schemes to reduce the transmitted power and improve the throughput of Bluetooth networks over reliable channels. The proposition depends on the Channel Quality Driven Data Rate (CQDDR) rules, which determine the suitable packet for the transmission process according to the channel conditions. The proposed packet formats are studied over Additive White Gaussian Noise (AWGN) and fading channels.

In Bluetooth systems, there are many types of packets that can be transmitted according to channel conditions. For good channel conditions, packets of long size are used for transmission and this will increase the throughput of wireless communication system. On the other hand, in bad channel conditions, packets of small size are used in transmission and the throughput of system will be decreased [1].

In fact, the Bluetooth employs variable-size packets. These packets occupy different numbers of time-slots up to a maximum of five slots; each time-slot length is 625 μs. Bluetooth V. 2.1 has brought EDR packet types [2]. These EDR packets support gross air rates of 2 Mbps and 3 Mbps through Л/4-DQPSK and 8 DPSK

M.A.M. El-Bendary, *Developing Security Tools of WSN and WBAN Networks Applications*, Lecture Notes in Electrical Engineering 316, DOI 10.1007/978-4-431-55069-3_4

modulation schemes, respectively. Different variants of these EDR packets will be discussed in this chapter. So, according to the truth that the Bluetooth is variable size packets technology, it employed the adaptive packets as discussed in Chap. 2. This chapter presents some of proposed different packets length to cover the gap between the short and long packets length performance.

The chapter is organized as follows. In Sect. 4.2, the related work to packet formats is discussed. In Sect. 4.3, the proposed modifications are presented. In Sect. 4.4, the Packet Error Probability (PEP) is presented. The simulation assumptions are given in Sect. 4.5. The simulation results are introduced in Sect. 4.6. Finally, the chapter is concluded in Sect. 4.7.

4.2 Related Work

Several authors have analyzed the performance of the Bluetooth technology utilization for implementing the WSN. Also, the IEEE 802.15.4 protocol [3] has been designed specifically for short-range communications in low-power sensor networks and is supported by most academic and commercial sensor nodes. The classical Bluetooth basic data rate versions uses the error control schemes the expurgated Hamming (15, 10) code used in the Bluetooth standard [4]. The most appreciable work in the coding of the payload field and EDR packets was introduced in [5]. In [6], other error control codes for improving Bluetooth performance were proposed. They improved the performance but reduced the Payload (PL) field length with an increase in complexity. The propositions of FEC bearing Data Medium (DM) packets for EDR were proposed in [7]. Also, in [8], the author proposes Reed-Solomon (RS) cods to enhance the Bluetooth reliability. In the same manner, all proposed cases improve the performance, but reduce the throughput, increase the complexity, and consume the power.

This chapter proposes a simple modification of the packet contents in good channel conditions. Also, it enforces the capability of the Bluetooth system of employing packets with different sizes through proposition a new adaptive packet format. These proposed formats have been chosen according to CQDDR rules with a more flexibility in the choice of the packet type.

Also, the performance of classic and EDR Bluetooth packets with standard and proposed formats is investigated. In addition, we activate the Received Signal Strength Indicator (RSSI) role for improving the throughput. The proposition is applied to classic and EDR Bluetooth packets.

4.3 Throughput Definition of WSN and WBAN

In general the throughput of the communications system can be defined as the amount of data transferred from the source to the destination or processed in a specific amount of time. It is also defined as the average number of information bits per second that are transmitted, successfully. It can be expressed as follows:

$$T = \frac{I(1 - p_J)}{T_s\left(N_1 + \sum_{j=2}^{J} N_j p_{j+1}\right)} \tag{4.1}$$

where T_s is the symbol period, and p_j is the probability of decoding failure in transmissions 1 through j. That is,

$$\begin{aligned} p_j &= P\left[\bigcap_{k=1}^{j} \text{Decoding failure in the } kth \text{ transmission}\right] \\ &= P[\text{Decoding failure in the } jth \text{ transmission}] = \mathrm{BLER_j} \end{aligned}$$

where the BLER denotes the block error rate achieved after the jth transmission.

The average packet delay, δ, is the expected time that elapses from the moment the packet is first transmitted over the channel to the moment the packet is successfully decoded. It is given by [9]:

$$\delta = T_{trans} + T_{wait} = T_s \frac{N_1 + \sum_{j=2}^{J} N_j p_{j-1}}{1 - p_J} + \frac{T_r \sum_{j=2}^{J} p_j}{1 - p_J} \tag{4.2}$$

where, T_{wait} is the average cumulative time between transmissions of a packet until successful decoding, and T_r is the time interval between transmissions.

Reducing the number of retransmissions can increase throughput. Also, increasing the number of retransmission leads to higher packet delays [10].

The throughput is the essence of what looks Designers of wireless networks in general, and it is a tangible product of the performance efficiency of the communication system. It measures the success of its effectiveness. With assuming that the safety factor on the wireless networks is absolute, the throughput remains the true measure of the system working efficiency.

In general, the throughput is defined as the amount of success received data at a time. Also, the user throughput is defined as the average number of information bits per second transmitted successfully.

So, the throughput of WSN network is defined by the area which contains several sensing station. The concept of area throughput is introduced in [14], it is defined as that the amount of data per second successfully transmitted to the sinks from a given

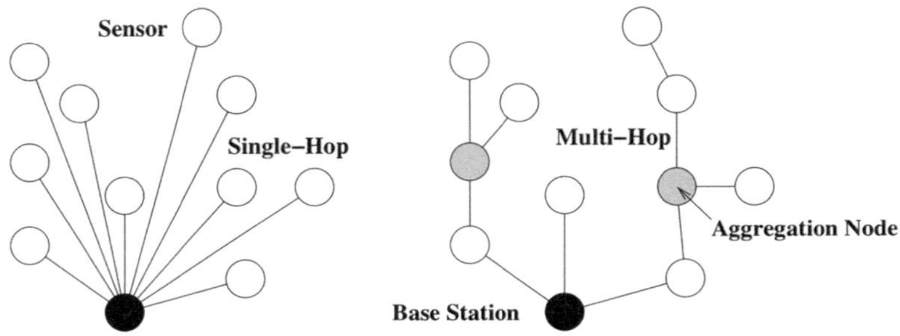

Fig. 4.1 Different scenarios of WSN topology with throughput concept

area. The previous definition can be true with the concept of a sensing nods group serves the same area as shown the Fig. 4.1

With respect to the cooperative work between nodes is not found. Every node talks about itself. So, the idea of area throughput is not reasonable to launch this term on all the network applications.

So, the definition can be classified to more than one according to the view angle, we are looking to the monitored area in general, to the success node for capture event forwarded to the sink, or to the collected data at the sink. The throughput term for the WSN concept may be related to the types of topology and the way of nodes deployment. Also, there is sub-division section within the service area or not.

With considering the concept of coverage area of every sensing station 'node', the term of area throughput is not enough to evaluate the reliability of the WSN link, especially if there is no cooperatively working between the notes.

The general equation of throughput calculation is formulated in Eq. 4.1. From Eq. 4.1, there are many factors affects on the amount of throughput. The application and the topology of WSN utilization determine the throughput definition. That means the ED applications and the S.

In the application of wireless networks in the sensing fields, WSN and WBAN, almost there are several of wireless sensing stations (Sensing Node (SN)). The sensing process means monitoring specific area and transferring its change (i.e. earth quarter, forest fires (WSN example) or bio-function of human body wireless monitoring i.e. wireless ECG Electrical Cardio Graph). All of the mentioned example, there are more than one sensor, limited or huge number of sensing stations. The main job of the sensors network is the truth transferring the around environment to the sink or the controller. So, the real or the best express of throughput is said Area Throughput (AT), why?, the area throughput is the term do not joined with the number of sensors in this area.

The area here refers to the service area or the monitored area, it may be defined as a piece of land, forest, country boundary, agriculture field or human body, it is determined by the field of applications.

In the following the analysis of the throughput of sensors networks is discussed with unlimited number of sensors (SN).

The channel between the SN and the sink or controller can be considered one way channel. The SN sends unchanged packet format, the controller or the sink send changeable packet according to the type of SN feature (fixed SN or mobile can cooperate with adjacent SN or not). Can SN interact with the sink or it just sensing station, send the around environmental changes only in happened time.

Let j is the transmission index,

The code rate R_j attained by combing the transmissions is given by:

$$R_j = \frac{I}{\sum_{k=1}^{j} N_k}$$

The throughput of single sensor station (SN) per second in bits is given by the following equation:

$$\eta = \frac{I}{T_{trans.}} \tag{4.3}$$

$$T_{trans.} = T_s\left(N_1 + \sum_{j=2}^{J} N_j p_{j-1}\right)$$

$T_{trans.}$ is the average cumulative time duration of the packet transmissions until successful decoding, and Ts is the symbol period.

Bluetooth Packet Format

The standard packet format for old versions of Bluetooth is shown in Fig. 4.2, while the format of EDR Bluetooth packets is given in Fig. 4.3. The Bluetooth Packet contains three main fields AC, HD, and Payload field. The function of the access code (AC) is to identify the packets exchanged within a piconet, where each piconet has a unique access code. The access code is used to synchronize the slaves in a piconet to their master. The main function of the header (HD) of the Bluetooth

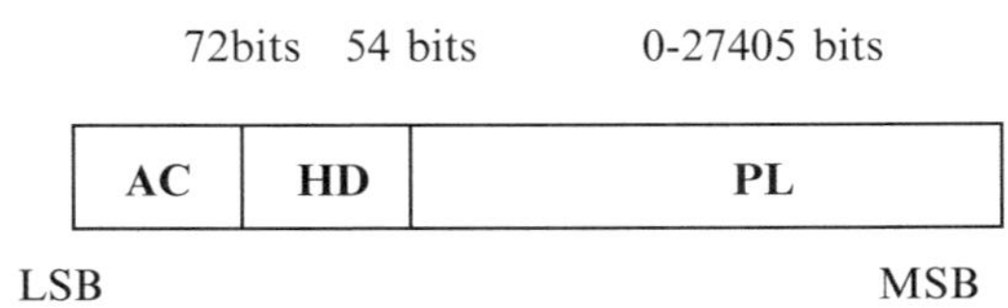

Fig. 4.2 Bluetooth classic packets format

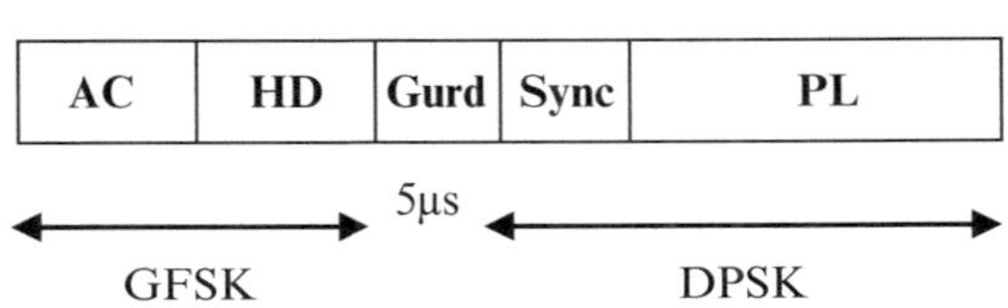

Fig. 4.3 Bluetooth EDR packet format

packet is to determine an individual slave address in the piconet by Logical Transport-Address (LT_ADDR) [9].

The last field of the Bluetooth packet is the payload. The functions of AC and HD are not changed in EDR packets. We focus in our study on a certain type of Bluetooth packets called ACL packets which refers to Asynchronous connectionless. Packets of the ACL payload may be one of two types; DM + and DHx. The M refers to medium data rate packets, while H refers to high data rate packets. The symbol x denotes the number of time slots between two hops in the frequency hopping system used [9]. It takes value 1, 3, and 5 referring to 1, 3, or 5 time slots between consecutive frequency hops. Always DMx packets are coded packets and DHx packets are uncoded packets [10].

4.4 Analysis of PEP

This section presents an analysis for the standard and proposed packet formats. Also, the effect of the proposed scheme on the system performance is studied. The throughput performance is affected by the Packet Error Probability (PEP) which is related to the packet size as follows [9].

$$T = \frac{N_i(1 - PEP)}{T_{trans}} \tag{4.4}$$

$$PEP = 1 - (1 - P_b)^L \tag{4.5}$$

where P_b is Bit Error Probability (BEP), and L is the number of bits in the packet. Equations 4.5 and 4.6 give the P_b over AWGN and Rayleigh fading channels, respectively, with Binary Phase Shift Keying (BPSK) modulation [10].

$$P_b = \frac{1}{2}\left(1 - \frac{1}{\sqrt{1 + \frac{1}{\sqrt{E_b/N_o}}}}\right) \tag{4.6}$$

$$P_b = \frac{1}{2} erfc\left(\sqrt{E_b/N_o}\right) \tag{4.7}$$

The PEP is decreased, when the packet size goes less. Equation 4.4 gives the PEP of uncoded packets. For coded packets, the PEP equation is given in [10].

$$PEP_c = 1 - (1 - P_c)^{N_c} \tag{4.8}$$

PEP_c is the packet error probability of coded packet, P_c is the codeword error probability, and N_c is the number of codewords in the packet.

The codeword error probability is a function of P_b, the number of correctable errors (t), and the length of codeword N_L. It can be expressed as:

$$P_c = \sum_{j=t+1}^{N_L} \binom{N_L}{j} P_b^j (1 - P_b)^{N_L - j} \tag{4.9}$$

The Bluetooth packet contains three main fields; the AC, the HD, and the PL. So the PEP of Bluetooth packets in general (coded and uncoded) is given as follows:

$$PEP_{bt} = 1 - (1 - P_{AC})(1 - P_{HD})(1 - P_{PL}) \tag{4.10}$$

In Eq. 4.10, the P_{AC} is the AC error probability, the P_{HD} is the HD error probability, and P_{PL} is the PL error probability. The first two fields, AC and HD are encoded by BCH (64, 30) code and repetition (3, 1) code, respectively. P_{AC} and P_{HD} are given as.

$$P_{AC} = \sum_{j=7}^{64} \binom{64}{j} P_b^j (1 - P_b)^{64-j} \tag{4.11}$$

$$P_{HD_w} = \sum_{j=2}^{3} \binom{3}{j} P_b^j (1 - P_b)^{3-j} \tag{4.12}$$

$$P_{HD} = 1 - (1 - P_{HD_w})^{18} \tag{4.13}$$

Classic and coded Bluetooth packets are encoded by the expurgated Hamming code (15, 10). So the codeword error probability can be expressed as:

$$P_{PL_w} = \sum_{j=2}^{15} \binom{15}{j} P_b^j (1 - P_b)^{15-j} \tag{4.14}$$

Then, the probability of encoded payloads is given as follows:

$$P_{PL} = 1 - (1 - P_{PL_w})^m \tag{4.15}$$

The number of codewords is m, m = 16, 100, and 183, for the DM_1, DM_3, and DM_5 packets, respectively.

m = 32 for $2DM_1$, m = 199 for $2DM_3$, and m = 365 for $2DM_5$. m = 47 for $3DM_1$, m = 297 for $3DM_3$, and m = 548 for $3DM_5$.

$$P_{bt_c} = 1 - (1 - P_{AC})(1 - P_{HD})(1 - P_{PL}) \tag{4.16}$$

For uncoded DH packets, the payloads are transmitted without FEC. The PEP of these packets is given as follows:

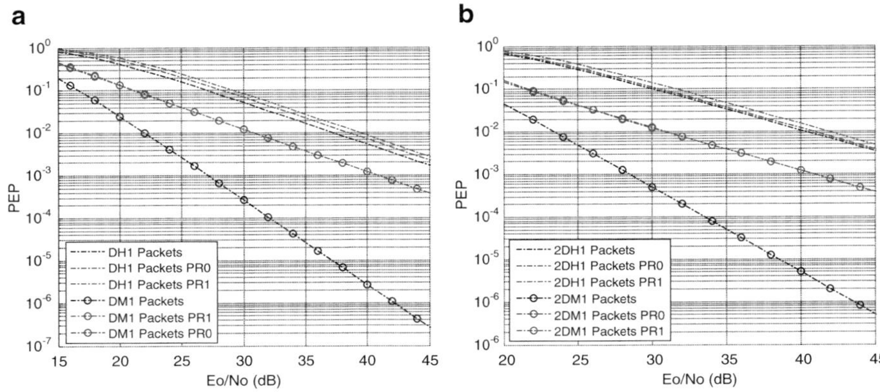

Fig. 4.4 PEP vs. E_b/N_o for over a fading channel. (**a**) For DM_1 and DH_1 of the standard and the proposed classic packets. (**b**) $2DM_1$ and $2DH_1$ of the standard and the proposed EDR packets

$$P_{PL_{unc}} = 1 - (1 - P_b)^{Lu} \tag{4.17}$$

L_u is the length of the uncoded payloads, L = 240, 1,500, 2,745, for the DH_1, DH_3, and DH_5 packets, respectively. m = 464 for 2-DH_1, m = 2,968 for 2-DH_3, and m = 5,464 for 2DH_5. m = 696 for 3-DH_1, m = 4,448 for 3DH_1, and m = 8,200 for 3DH_5. The uncoded classic Bluetooth packets error probability $P_{bt_{unc}}$ can be expressed as:

$$P_{bt_{unc}} = 1 - (1 - P_{AC})(1 - P_{HD})(1 - P_{PL_{unc}}) \tag{4.18}$$

The proposed scheme is studied analytically as shown in Fig. 4.4. This figure shows PEP variation for the Bluetooth packets with the standard and the proposed formats over a Rayleigh fading channel for classic and EDR packets. Equation 4.18 gives the Throughput (T) over Bluetooth network [11].

$$T = \frac{PL(1 - PEP)}{[T_s = (x + 1)t_S]} \tag{4.19}$$

PL is the user payload length, x is the number of time slots occupied by the Bluetooth packet, and t_s is the duration of the Bluetooth time slots. The T is used for evaluating the proposed schemes effectiveness as shown in Sect. 4.6.

4.5 The Proposed Scenarios

In this chapter, we study throughput enhancement over Bluetooth system with different types of packets such as EDR packets, classic packets, and the proposed packet formats over AWGN and Rayleigh flat-fading channels. This section

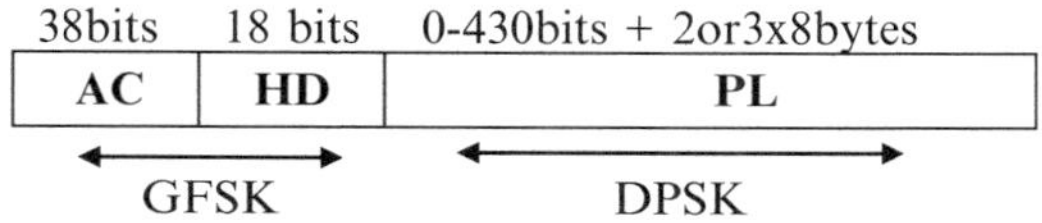

Fig. 4.5 Adaptive Bluetooth EDR packet format for $2DH_1$ EDR packets

proposes the use of different packet sizes. In Bluetooth packets, the AC and HD fields are encoded by the BCH (30, 64) code and the repetition (1, 3) code, respectively. These two fields are very important for establishing a Bluetooth link. So, they must be protected by FEC [9]. The length of the redundant fields in the AC and HD is more than 8 bytes. In good channel conditions, these redundant bits can be reserved to extend the PL, and this leads to an enhancement of the system throughput. In the proposed packet format, the length of Payload (PL) field can be extended to enhance throughput.

In the Bluetooth system, the packet size is determined in CQDDR through the RSSI, which senses the wireless link and channel conditions [10]. Our proposed packet formats are efficient in good channel conditions. The proposed adaptive formats are applied to classic and EDR Bluetooth packets. In EDR packets, we take the three main fields; AC, HD, and PL only into consideration. The proposed adaptive Bluetooth EDR packet format is shown in Fig. 4.5. The proposition is built on two real factors. First, DH packets are chosen by RSSI in the case of good channels. Second, the most discarded packets are due to uncoded PL [12].

In general, the transmitted packets of the Bluetooth are variable according to the channel conditions. So, a simple approach for enhancing the throughput is presented in this work. In good channel conditions, the redundant bits reduce the bandwidth efficiency. So, the coding rate (R_c) of the Bluetooth packet in standard form is given by $R_c = N_i/N_b$, where N_i denotes the body data length and N_b refers to the packet size, which contains the redundant data with length (N_r). With the application of the proposed approach, the coding rate is given by $R_c = (N_i + N_r)/(N_b - N_r)$. As shown in [10], the data Throughput (T) can be expressed by $T = (1 - PER)/D_{R_c}$, where PER is the packet error rate and D_{R_c} is the data rate. It can be formulated by $D_{R_c} = N_i/T_{trans}$, where T_{trans} refers to the time required for transmitting one packet [9].

4.6 Simulation Assumptions

The Monte Carlo simulation method is used in the simulation experiments to compare between the Bluetooth standard with the traditional expurgated Hamming (15, 10) code, EDR packets, and the proposed packets. This method ensures obtaining correct statistical results [10].

Hard decision is assumed at the receiver in the decoding process for all channel conditions. In the simulation, the interference effects are neglected. The modulation in our simulations is BPSK modulation scheme [11]. In some simulation experiments, a block-fading channel is assumed. It is a slow and frequency non-selective channel, where symbols in a block undergo a constant fading effect.

We will concentrate in our experiments on DH_1, DM_1 (classic Bluetooth packets), $2DH_1$, and $2DM_1$ (EDR Bluetooth packets), which are short packets, and $2DM_5$, $2DH_5$ as longer Bluetooth packets. MATLAB was used for carrying our simulation experiments of different cases [10].

4.7 Simulation Results

Several computer simulation experiments have been carried to test the performance of the classic and proposed packets. DH_1, DM_1, $2DH_1$, and $2DM_1$ packets have been considered in the computer simulations. Four scenarios have been applied to these packets, which are standard uncoded packets, standard encoded packets using Hamming code, Pr1 packets, and Pr2 packets.

Pr1: The AC and HD fields are transmitted without FEC.
Pr2: the redundant bits, which are reserved in the standard are added to the PL field.

These scenarios are applied to classic and EDR packets. Several experiments have been carried out for the purpose of comparison between the standard schemes and the proposed EDR packets. Also, the Number of Lost packets (NLP) and the Packet Error Rate (PER) have been studied with the variation of the channel Signal to Noise Ratio (SNR) [13].

4.7.1 Simulation Experiments with Short Packets

This section is devoted to evaluate and measure the Number of Lost Packets (NLP) and the transmission performance of the proposed schemes using DH_1, $2DH_1$, DM_1, and $2DM_1$ packets with the different scenarios over an AWGN and Rayleigh fading channels. In the first computer simulation experiment, theses packets have been transmitted over an AWGN channel with different SNR values. Table 4.1 tabulates the NPL variation with the channel SNR using different scenarios for the EDR packets over an AWGN channel.

An experiment has been carried out using short classic packets with DM_1, DM_1-Pr1, DH_1, DH_1- Pr1, and DH_1- Pr2 formats. The results of this experiment are shown in Fig. 4.6. This figure shows the variation of the PER of the standard and the proposed EDR packets with channel SNR over an AWGN channel. Simulation results reveal that with the enhancement of channel condition, the proposed packet formats can be used to increase the amount of transmitted data over Bluetooth networks. Also, the proposed schemes can be used as available choices with the standard packets.

Table 4.1 NLP for EDR packets over an AWGN channel with different scenarios

SNR	NLP for EDR packets over an AWGN channel					
	AC&HD	AC&HD NO FEC	$2DM_1$	$2DH_1$	$2DH_1$ Pr_1	$2DH_1$ Pr_2
0 dB	4,317	9,810	ALL	ALL	ALL	ALL
2 dB	816	8,389	9,504	ALL	ALL	ALL
4 dB	85	4,530	3,381	9,959	9,979	9,989
6 dB	2	1,106	151	6,383	6,844	7,261
8 dB	0	97	2	790	840	955
10 dB	0	1	0	16	23	27

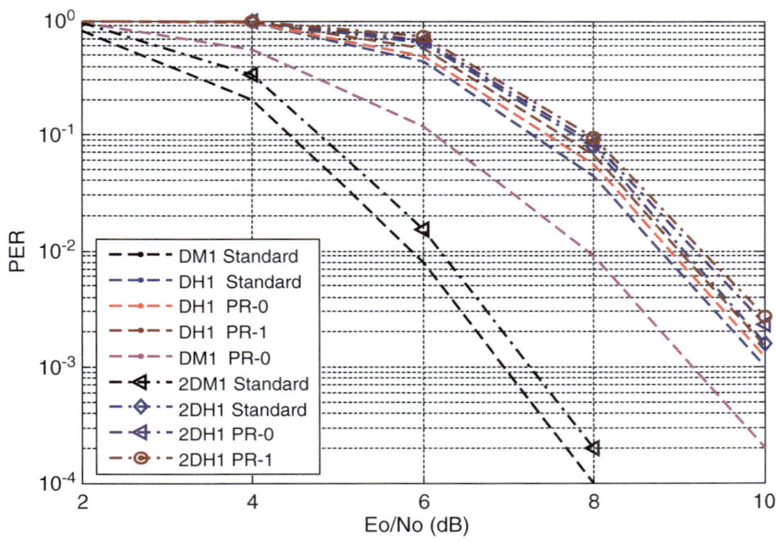

Fig. 4.6 PER vs. E_b/N_o over an AWGN channel with different scenarios using DM_1, DH_1, $2DM_1$ and $2DH_1$ of the standard and the proposed classic and EDR packets for data transmission

The performance of the standard and the proposed EDR packets over a Rayleigh fading channel is shown in Fig. 4.7. This figure shows the variation of the PER with the channel SNR. The NLP of the standard and the proposed EDR packets over a fading channel is tabulated in Table 4.2. As shown in these results, the proposed packet format can be accepted with the channel conditions improvement.

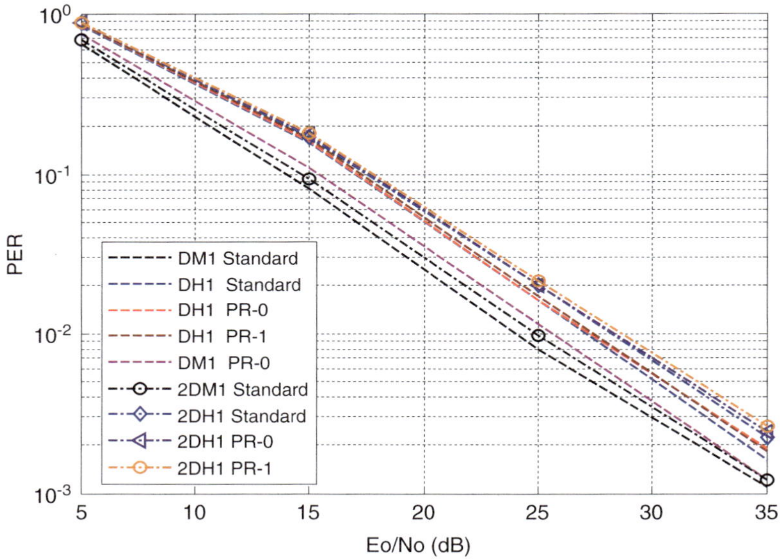

Fig. 4.7 PER vs. E_b/N_o over a fading channel with different scenarios using DM_1, DH_1, $2DM_1$ and $2DH_1$ of the standard and the proposed classic and EDR packets for data transmission

Table 4.2 NLP for EDR packet over a fading channel with different scenarios

SNR	NLP for EDR packets over a fading channel					
	AC&HD	AC&HD NO FEC	$2DM_1$	$2DH_1$	$2DH_1$ Pr_1	$2DH_1$ Pr_2
5 dB	3,672	6,952	6,879	8,718	8,781	8,834
15 dB	423	1,056	923	1,715	1,778	1,803
25 dB	37	109	98	202	201	214
35 dB	4	14	12	22	24	26

4.7.2 Long Packets Performance Over a Fading Channel

This section is devoted for studying the proposed formats on long Bluetooth packets. The computer simulation experiments have been carried out using DH_5, $2DH_5$, DM_5, and $2DM_5$ packets with different scenarios over Rayleigh fading channels. In the first experiment, the long EDR packets are transmitted over a

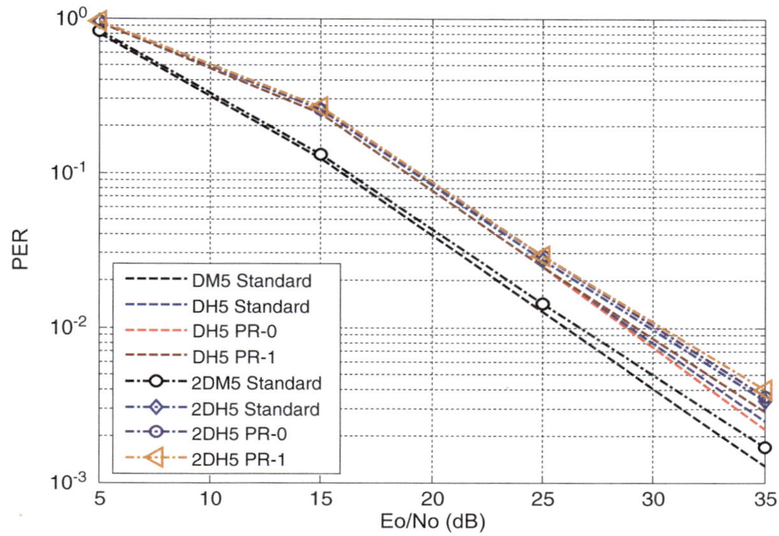

Fig. 4.8 PER vs. E_b/N_o over a fading channel with different scenarios using DM_5, DH_5, $2DM_5$, and $2DH_5$ of the standard and the proposed classic and EDR packets for data transmission

fading channel with different SNR values. The results of this experiment are shown in Fig. 4.8. This figure gives the PER of the standard and proposed long classic and EDR packets over Rayleigh fading channels with the variation of the SNR.

Our simulation results reveal that the PL of EDR and classic Bluetooth packets can be increased by 8, 16, or 24 bytes according to the packet type. The proposed packets are efficient for shorter packets; DH_1 and $2DH_1$.

4.7.3 General Comparison

This section gives a comparison between the standard and the proposed packets for data transmission with different types of packets based on the throughput (T). Figures 4.9 and 4.10 show the variation of the throughput with the channel SNR for the data transmission over an AWGN channel for the short classic and the short EDR packets, respectively. These figures indicate the efficiency of the proposed packet format for the short packets. As shown in these figures, the proposed adaptive packets provide higher throughput with good channel conditions. Figures 4.11 and 4.12 give the throughput variation for the data transmission over a fading channel for classic and EDR packets.

The obtained results reveal that for SNR > 6 dB, the proposed packets on Pr-1 basis give a high throughput for short packets over an AWGN channel.

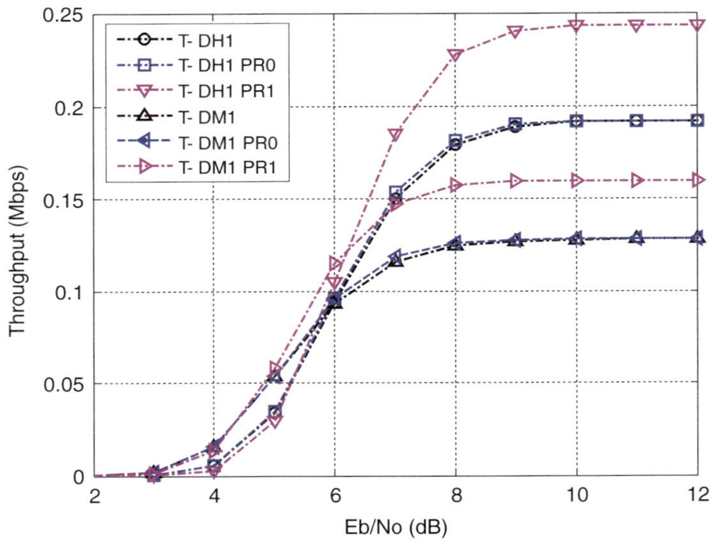

Fig. 4.9 Throughput vs. E_b/N_o for the data transmission over an AWGN channel using DM_1 and DH_1 packets with different proposed scenarios

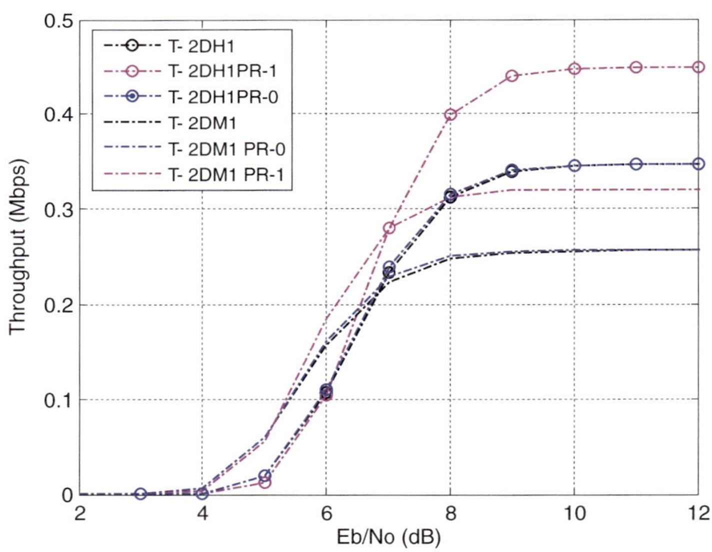

Fig. 4.10 Throughput vs. E_b/N_o for the data transmission over an AWGN channel using $2DM_1$ and $2DH_1$ EDR packets with different proposed scenarios

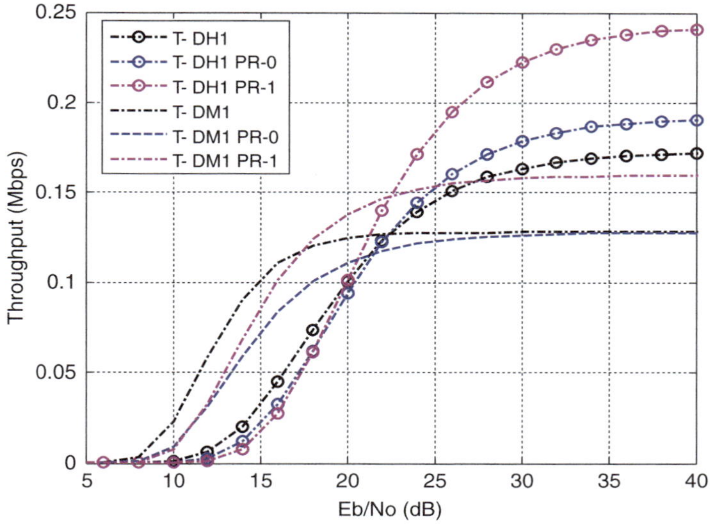

Fig. 4.11 Throughput vs. E_b/N_o for the data transmission over a fading channel using DM_1 and DH_1 classic packets with different proposed scenarios

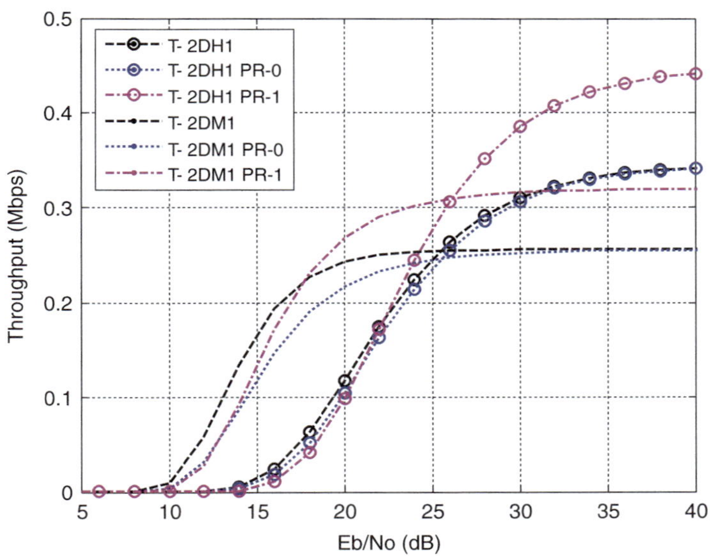

Fig. 4.12 Throughput vs. E_b/N_o for the data transmission over a fading channel using $2DM_1$ and $2DH_1$ EDR packets with different proposed scenarios

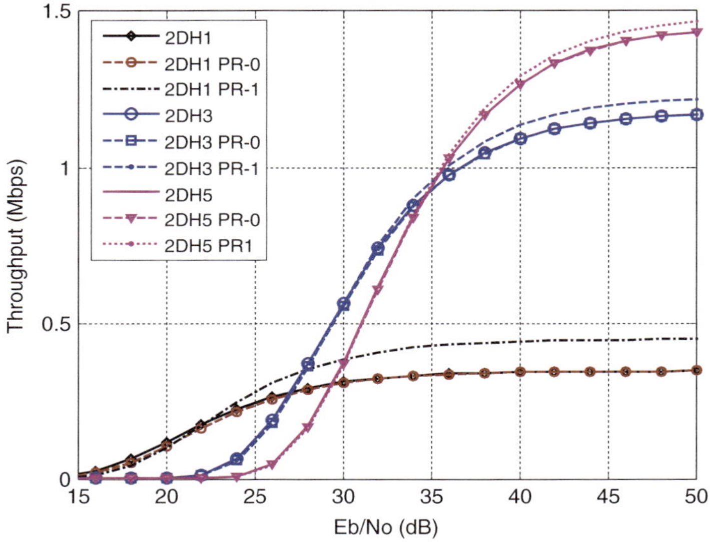

Fig. 4.13 Throughput vs. E_b/N_o over a fading channel using $2DH_x$ packets with different proposed scenarios (analytical study)

The standard packets produce high throughput at SNR < 6 dB. On other hand, over a fading channel, the proposed schemes give a high throughput at SNR > 17 dB.

From the previous results, the proposed scenarios are suitable with short packet. Also, the proposed packets are controlled by the channel conditions through the CQDDR rules. The following figures indicate the proposed scenarios for long packets. Equation 4.16 is used to calculate the throughput of the standard and proposed packets. As shown in the following figures, the proposed scenarios give higher enhanced throughput with the EDR packets. The results covered the relation between the throughput and the SNR with the $2DH_x$ and $2DM_x$ packets over the fading channels. Also, the $3DH_x$ and $3DM_x$ packets over the fading channel with the standard and the proposed scenarios have been studied. Figure 4.13 gives the relation between the throughput and the E_b/N_o in the case of $2DH_x$ packets with the proposed format. Figure 4.14 indicates the throughput variation with the BEP over the fading channel.

The following Figs. 4.15 and 4.16 give the throughput variation with E_b/N_o and BEP, respectively. These figures reveal that the amount of successfully transmitted data can be increased with proposed scenarios under good channel conditions.

Figures 4.17 and 4.18 gives the throughput variation with E_b/N_o and BEP, respectively for the $3DH_x$ packets. These figures indicate that the long packets are not suitable for the proposed scenarios; the EDR packets support the proposed format.

Figures 4.19 and 4.20 gives the throughput variation with the variety of E_b/N_o and BEP, respectively for the $3DM_x$ packets. These figures indicate that the long

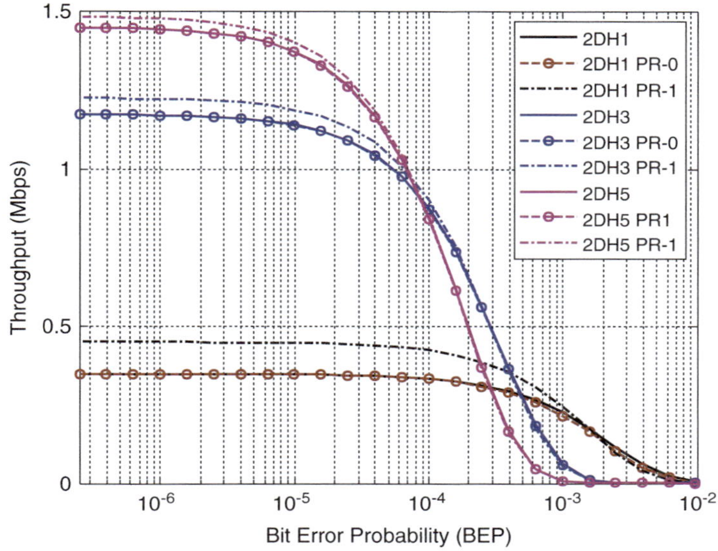

Fig. 4.14 Throughput vs. BEP over a fading channel using $2DH_x$ packets with different proposed scenarios (analytical study)

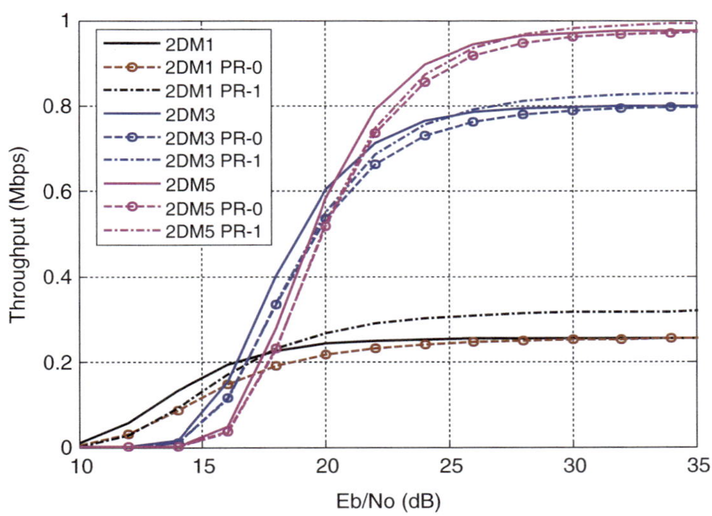

Fig. 4.15 Throughput vs. E_b/N_o over a fading channel using $2DM_x$ packets with different proposed scenarios (analytical study)

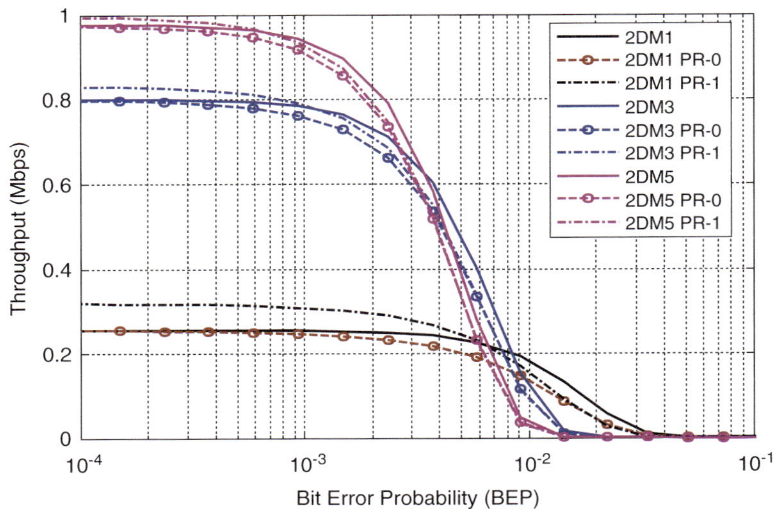

Fig. 4.16 Throughput vs. BEP over a fading channel using $2DM_x$ packets with different proposed scenarios (analytical study)

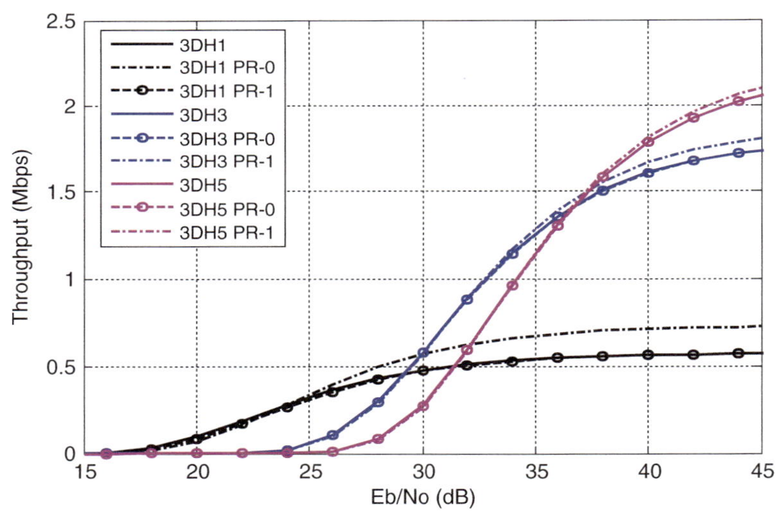

Fig. 4.17 Throughput vs. E_b/N_o over a fading channel using $3DH_x$ packets with different proposed scenarios (analytical study)

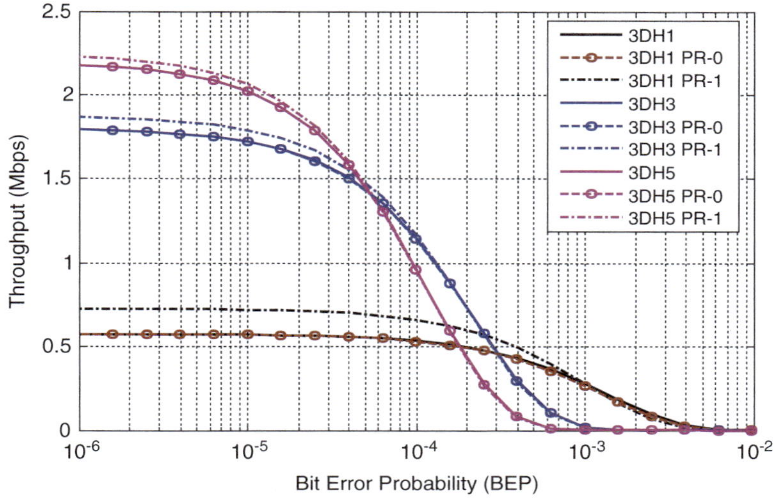

Fig. 4.18 Throughput vs. BEP over a fading channel using $3DH_x$ packets with different proposed scenarios (analytical study)

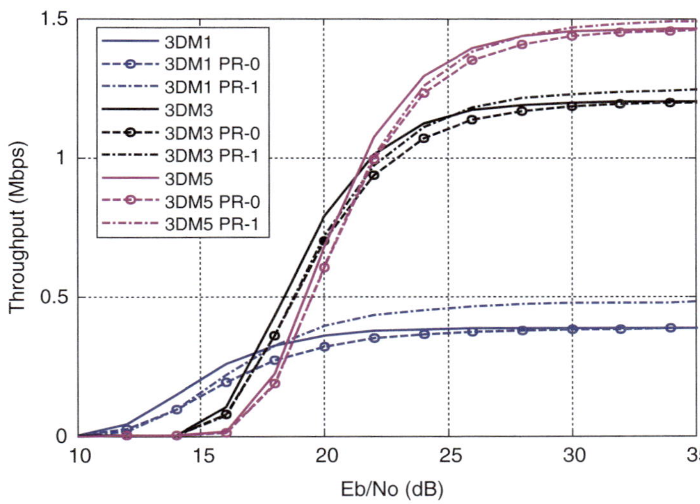

Fig. 4.19 Throughput vs. E_b/N_o over a fading channel using $3DM_x$ packets with different proposed scenarios (analytically studying)

packets are not suitable for the proposed scenarios also; the EDR packets support the proposed format.

Figures 4.21 and 4.22 give the throughput variation with E_b/N_o and BEP, respectively for the DH_1, $2DH_1$, and $3DH_1$ packets. These figures indicate that the shortest uncoded EDR packets are very suitable for the proposed scenarios also; the EDR packets support the proposed format.

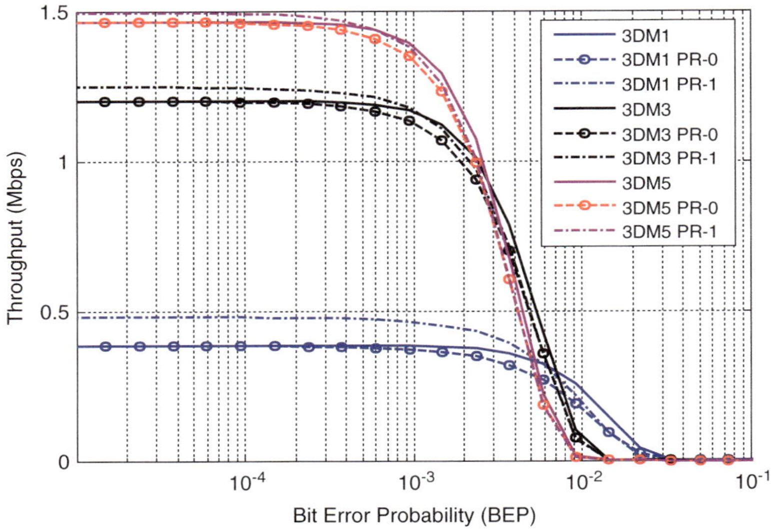

Fig. 4.20 Throughput vs. BEP over a fading channel using $3DM_x$ packets with different proposed scenarios (analytically studying)

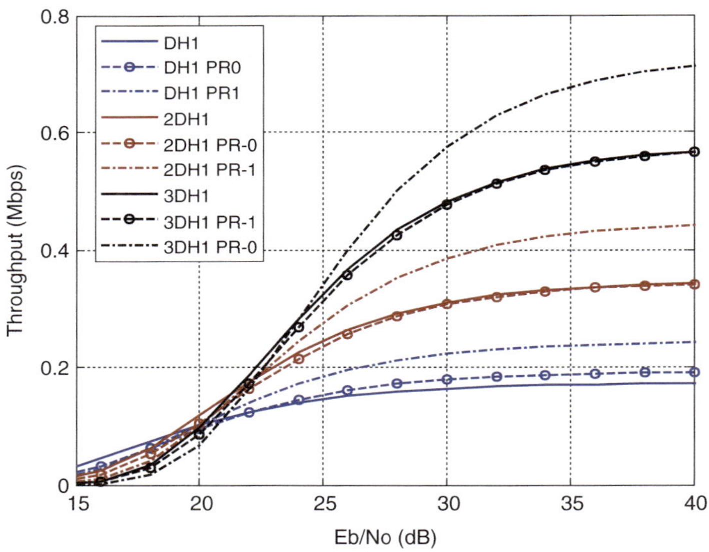

Fig. 4.21 Throughput vs. E_b/N_o over a fading channel using (1, 2, and $3DH_1$) packets with different proposed scenarios (analytical study)

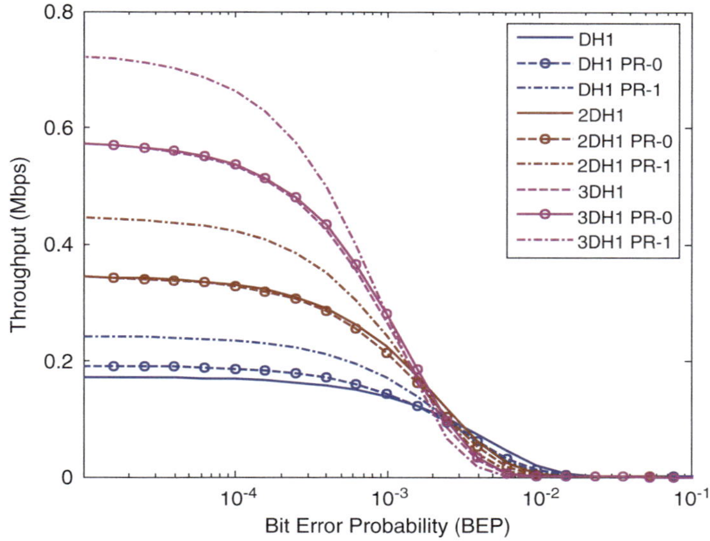

Fig. 4.22 Throughput vs. BEP over a fading channel using (1, 2, and $3DH_1$) packets with different proposed scenarios (analytical study)

4.8 Chapter Recommendation

We have proposed a group of adaptive packets with different formats depending on the CQDDR concepts for improving the performance of data transmission over Bluetooth networks. The proposed idea improved the throughput of the data transmission over classic and EDR packets using different proposed schemes. Simulation experiments revealed that the proposed formats are good for the short packets with good channel conditions. The CQDDR is employed for selecting the transmitted packet size by the RSSI according to the channel conditions.

References

1. Galli S, Famolari D, Kodama T (2004) Bluetooth: channel coding considerations. In: IEEE 59th vehicular technology conference, pp 2605–2609
2. Golmie N, Van Dck RE, Soltanian A (2001) Interference of Bluetooth and IEEE 802.11: simulation modeling and performance evaluation. In: Proceedings ACM international workshop on modeling, analysis, and simulation of wireless and mobile systems, Italy
3. Gutierrez JA, Naeve M, Callaway E, Bourgeois M, Mitter V, Heile B (2001) A developing standard for low-power low-cost wireless personal area networks. IEEE Netw 15(5):12–19
4. Mohamed MAM, Abou El-Azm A, El-Fishwy N, El-Tokhy MAR, Abd El-Samie FE (2008) Optimization of Bluetooth packet format for efficient performance. Progr Electromagn Res M 1:101–110

5. Meghanathan N (2009) Impact of range of simulation time and network shape on the hop count and stability of routes in mobile ad hoc networks. IENG Int J Comput Sci. Advanced online publication, 36:1–01
6. Mohamed MAM, Abou El-Azm A, El-Fishawy NA, El-Tokhy MAR, Shawky F, Abd El-Samie FE (2008) Bluetooth performance improvement using convolutional codes. In: Proceedings of the 2nd international conference on electrical systems design & technologies, Hammamet, 8–10 Nov 2008
7. Mohamed MAM, Abou El-Azm A, El-Fishawy NA, El-Tokhy MAR, Shawky F, Abd El-Samie FE (2008) Bluetooth performance improvement over different channels through channel coding. In: Proceedings of the 5th international multi-IEEE conference on systems, signals and devices, Jordan
8. Vafi S, Wysocki T (2005) Performance of convolutional interleavers with different spacing parameters in turbo codes. In: Proceedings of the 6th Australian workshop on communications theory. Brisbane, Qld pp 8–12
9. NguyenVD, Kuchenbecker H (2001) Block interleaving for soft decision viterbi decoding in OFDM systems. In: IEEE 54th VTC Fall, vol 1, pp 470–474
10. Shi YQ, Zhang XM, Ni Z-C, Ansari N (2004) Interleaving for combating error bursts. IEEE Circuit Syst Mag 4:29–42, First Quarter
11. Farag EN, Elmasry MI (1999) Mixed signal VLSI wireless design circuits and system, 1st edn. Kluwer Springer, Circuits and Systems
12. Vafi S, Wysocki TA (2006) Application of convolutional interleavers in turbo codes with unequal error protection. JTIT, J Telecommun Inf Technol, 1:17–23
13. Visotsky E, Sun Y, Tripathi V, Honig ML, Peterson R (2005) Reliability-based incremental redundancy with convolutional codes. IEEE Trans Commun 53(6):987–997
14. Chiara B, Flavio V, Roberto V (2009) Area throughput of an IEEE 802.15.4 based wireless sensor network. In: EWSN –09 Proceedings of the 6th European Conference on WSN, Springer-Verlag, Berlin, Heidelberg, pp 1–16

Chapter 5
WSN Security Needs

Abstract The secured link between the sink and the sensing nodes in the WSN is a vital factor in many applicable fields. There are many applications of this network concern the human life or the national security of the countries. In this chapter, the different threats of WSN are discussed.

Keywords WSN security • Threats • Ad-hoc threats • WPANs security • Security challenges

5.1 Introduction

Although the suitability of the wireless technologies for many applications and different fields, there is essential fact that the nature of wireless communication makes wireless networks more vulnerable to attacks than wired networks. Moreover, WSNs are often deployed in uncontrolled environments, which make it susceptible to physical tampering. The limited computational capabilities and energy resources are additional challenges that need to be dealt with in designing security scheme for WSNs. The following are the main threats on the integrity of the WSN data [1–3]:

Data Modification Attack An adversary modifies the value of one or more the data readings either by hijacking the sender sensor or inserting itself between the sender and receivers.

False Data Insertion An adversary can compromise existing nodes and inject a false message with false information. It is also possible that the adversary add new nodes to the sensor networks that feed false data. Such attack also consumes the energy resources of other sensor nodes.

Data deletion Data deletion attack can take place by dropping individual data readings or dropping one or more groups and preventing them from reaching to the intended recipient.

Denial of Service Denial of service attacks on a wireless sensor network may take on several forms, e.g., disrupting the radio link, misroute sensor data, or exhaust

M.A.M. El-Bendary, *Developing Security Tools of WSN and WBAN Networks Applications*, Lecture Notes in Electrical Engineering 316, DOI 10.1007/978-4-431-55069-3_5

node resources. Section 5.5 shows how the proposed technique deals with each of the above attacks.

5.2 Security Goals

The security goals of WSN are the same for any wireless networks and applications. The main difference for WSN situation, the security is very critical issue. The applications of WSN may be applied on very essential fields which directly affects on the country security or the economic development. The security techniques may be optional in some applications of WSN, on the other hand, the WSN is indispensable for many application such as the military fields, monitoring the radiation leakage of nuclear sites, the spread of epidemics and biological and chemical weapons, monitoring and control in the cultivation of vast areas agricultural, environmental monitoring, early detection of forest fire, and etc. In these fields of WSN application, there is no substitute for the use of high-efficiency techniques of security.

The Security is an important issue for ad hoc networks, especially for those security-sensitive applications. To secure an ad hoc network, we consider the following attributes: availability, confidentiality, integrity, authentication, and non-repudiation.

5.3 Fundamentals of Network Security

Computer and network security is the collection of all policies, mechanisms, and services that afford a computer system or network the required protection from unauthorized access or unintended uses. Most security mechanisms are built to address three well-known services in the CIA security model: Confidentiality, Integrity, and Availability. The following describes these services in more detail:

1. Confidentiality: Security mechanisms must ensure that only the intended receiver can correctly interpret a message and that unauthorized access and usage is prevented. For example, confidentiality ensures that sensitive information such as a person's social security number or credit card information are not obtained by an unauthorized individual.
2. Integrity: Security mechanisms must ensure that a message cannot be modified as it propagates from the sender to the receiver, that is, unauthorized individuals should not be able to destroy or alter the contents of sensitive information.
3. Availability: Security mechanisms must ensure that a system or network and its applications are able to perform their tasks at any time without interruption. Availability is often measured in terms of percentages of up or down time.

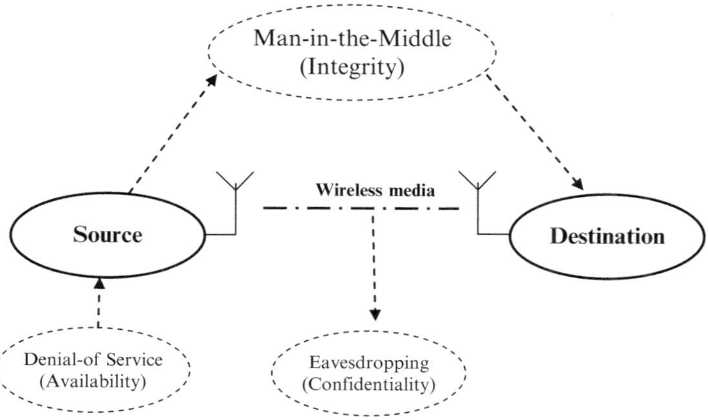

Fig. 5.1 Examples of attacks and the CIA model

Examples of attacks on a transmission between a sender and its intended receiver are shown in Fig. 5.1. The figure shows the *Eavesdropping, it* refers to the reception of a message by an unauthorized individual, which can be prevented using confidentiality measures. A *man-in-the-middle* attack refers to a situation where an unauthorized individual or system positions itself between the sender and receiver such that the sender's messages are intercepted, modified, and retransmitted to the receiver (where the receiver believes the received message came directly from the original sender). This illustrates the need for integrity mechanisms. Finally, a *denial-of-service* attack refers to an adversary's attempt to disrupt the transmission or service provided by the sender. For example, the adversary can overload the sender with requests and tasks such that the sender is not able to transmit its message (in a timely fashion) to the receiver. This type of attack necessitates security mechanisms that ensure availability.

In addition to the three components of the CIA triad, *authentication* refers to the process of establishing or confirming identify of a user or a device, ensuring that a message came from who it claims to have come from. Also, *non-repudiation* refers to the process of proving that a person or device has performed a transaction or transmission. *Digital signatures* are often used to support both authentication and non repudiation, but are also used to provide confidence that a message has not been altered (i.e., integrity). In all types of communication networks, there are several fundamental security mechanisms that can be used to provide confidentiality, integrity, and availability. *Cryptography* is the process of hiding and protecting information using encoding and decoding mechanisms. In *symmetric key cryptography*, a single key between two communicating parties is used for the encryption and decryption of a message.

The source represents the sensing station position and the destination in this model represents the sink "the WSN controller". In some of other WSN implementation scenarios, the destination may be represents another sensing node acts as head cluster.

5.3.1 Security Analysis

Security and privacy are enormous challenges in all types of wired and wireless networks. These challenges are of even greater importance in wireless sensor networks, where the unique characteristics of these networks and the application purposes they serve make them attractive targets for intrusions and other attacks. In applications such as battlefield surveillance and assessment, target tracking, monitoring civil infrastructure such as bridges and tunnels, and assessment of disaster zones to guide emergency response activities, any breach of security, compromise of information, or disruption of correct application behavior can have very serious consequences. Sensor networks are frequently used in remote areas, left to operate unattended and therefore providing an easy target for physical attacks, unauthorized access, and tampering. Sensor nodes are typically very resource-constrained and operate in harsh environments, which further facilitate compromises and makes it often difficult to distinguish security breaches from node failures, varying link qualities, and other commonly found challenges in sensor networks. Finally, these resource constraints require security mechanisms that are customized for WSN applications, such that the limited resources are used efficiently. This chapter provides an overview of the security concerns of WSNs and illustrates possible solutions to providing security and privacy protection. Note that the terms attacker, intruder, and adversary are used interchangeably to describe an entity (person or device) that performs an attack on a network or system.

5.3.2 Ad-Hoc Security

Some of researchers discuss the WSN as Ad-hoc networks. In this section, the need of security over the Ad-hoc networks is described. Some attacks scenarios also, are shown in Figs. 5.2 and 5.3. The Security in Mobile Ad-Hoc Network is the most important concern for the basic functionality of network. The availability of network services, confidentiality and integrity of the data can be achieved by assuring that security issues have been met. MANETs often suffer from security attacks because of its features like open medium, changing its topology dynamically, lack of central monitoring and management, cooperative algorithms and no clear defense mechanism. These factors have changed the battle field situation for the MANETs against the security threats [4].

The MANETs work without a centralized administration where the nodes communicate with each other on the basis of mutual trust. This characteristic makes MANETs more vulnerable to be exploited by an attacker inside the network.

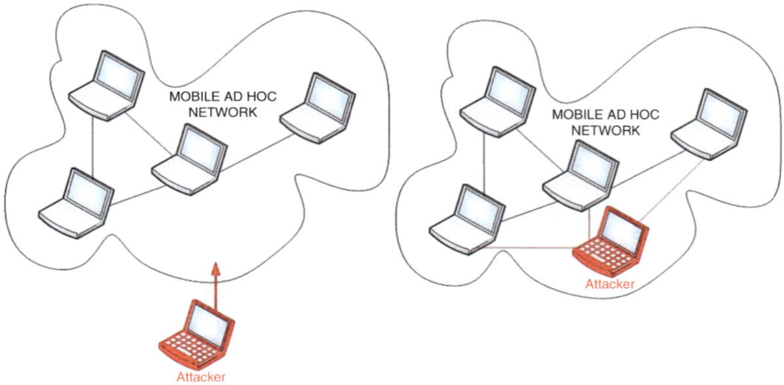

Fig. 5.2 External and internal attacks in MANETs [9]

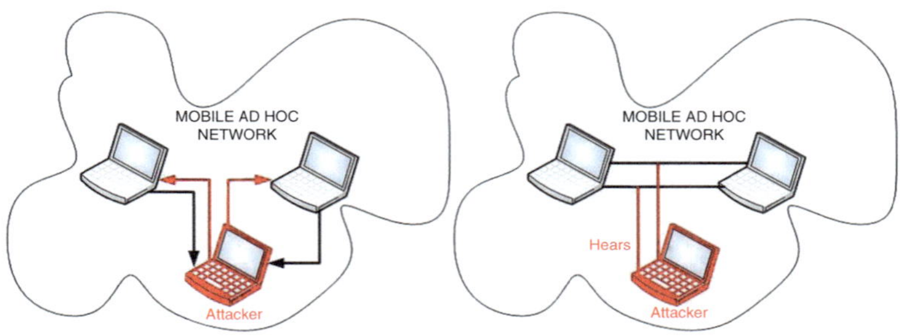

Fig. 5.3 Active and passive attack in MANETs

Wireless links also makes the MANETs more susceptible to attacks, which make it easier for the attacker to go inside the network and get access to the ongoing communication [5]. Mobile nodes present within the range of wireless link can overhear and even participate in the network.

MANETs must have a secure way for transmission and communication and this is a quite challenging and vital issue as there is increasing threats of attack on the Mobile Networks. Security is the cry of the day. In order to provide secure communication and transmission, the engineers must understand different types of attacks and their effects on the MANETs. Wormhole attack, Black hole attack, Sybil attack, flooding attack, routing table overflow attack, Denial of Service (DoS), selfish node misbehaving, impersonation attack are kind of attacks that a MANET can suffer from. A MANET is more open to these kinds of attacks because communication is based on mutual trust between the nodes, there is no central point for network management, no authorization facility, vigorously changing topology and limited resources.

5.4 Challenges of WSN Security

The challenges of WSN limit its application. The challenges of security utilization for WSN are not much different from it challenges of applications. Naturally, the security techniques provide complexity and in addition increasing the power consumption. The most of security algorithms require a significant amount of memory and storage spaces are therefore infeasible for such sensors, where the memory one of the computational complexity components.

WSNs exhibit a variety of unique challenges that must be considered when addressing the security concerns that may arise in sensor network applications:

1. *Resource constraints*: Traditional security mechanisms that have high overheads are not suitable for resource-constrained WSNs. Many security mechanisms are computationally expensive or require communication with other nodes or "remote" devices (e.g., for authorization purposes), thereby leading to energy overheads. Small sensor devices are also constrained in their available memory and storage capacities.
2. *Lack of central control*: It is often infeasible to have a central point of control in sensor networks, for example, because of their large scale, resource constraints, and network dynamics (topology changes, network partitioning). Therefore, security solutions should be decentralized and nodes must collaborate to achieve security.
3. *Remote location*: The first line of defense against security attacks is to provide only controlled physical access to a sensor node. Many WSNs are left unattended, because they are operated in remote and hard-to-reach locations, deployed in environments open to public access, or so large that it would be infeasible to continuously monitor and protect sensor nodes from attacks. These challenges make it difficult to prevent unauthorized physical access and to detect tampering with the sensor devices, particularly since the low cost of many sensor nodes may prohibit advanced (and expensive) protective measures.

   The distributed control of WSN and its service area division to small clusters area may be helpful to reduce the challenge of remote location. Every cluster of sensors has head nose manages the security requirements and there is unique secret key for the nodes of the cluster. So, the security management of the cluster is the head responsibility.
4. *Error-prone communication*: Packets in WSNs may be lost or corrupted due to a variety of reasons, including channel errors, routing failures, and collisions. This may interfere with some security mechanisms or their ability to obtain critical event reports. Furthermore, this may make it difficult to distinguish "benign" erroneous communications or node and link failures from malicious attacks.

5.4.1 Security Challenges and Nature of WSN

This section discusses the security need based on the characteristics of the WSN networks. Certain characteristics of sensor networks, on the other hand, facilitate the provision of security. For example, the self-managing and self-repairing nature of a WSN may allow it to continue to operate even if a sensor or entire regions of the sensor network have been compromised. Redundancy in a sensor network allows it to gather information about events in the environment even when some sensors are unavailable due to an attack. Furthermore, this redundancy can be used to detect, isolate, and mask potentially compromised nodes [6].

Data collected by sensors may contain sensitive information and should not be leaked to unauthorized devices. Further, encryption keys and information about sensors themselves (e.g., identity, location, etc.) must be protected to prevent eavesdropping and attacks based on traffic analysis. These challenges require measures that provide data confidentiality for sensor networks. Integrity is required to prevent adversaries from modifying sensor data, for example, with the purpose of injecting false readings and therefore affecting the response to the sensor readings. Authentication is necessary to ensure that any data disseminated in a sensor network originates from the correct source, particularly when a single node controls the entire network (e.g., a base station establishing routes or distributing multicast tree information) [7].

The security of WSN needs tools and techniques. In the truth, the security techniques are one process and stage in the communication system. The security process like the protection data schemes, some time it is very useful and with its absent the communication system fail, and another time it can be considered overload on the system. It needs power for processing the consumed power increased with the complexity of the security techniques increasing. Also, it loads the system extra time delay for the encryption before transmission and decryption after the receiving. The last point some security technique needs extra memory for carrying out its algorithm. The memory length plays role in the system complexity computing [8, 9].

5.5 WSN Implementation Technologies

There are some wireless technologies are utilized to implement the local WSN application. These technologies are WPANs technologies such as the ZigBee network and Bluetooth networks. In the following Fig. 5.4, the different wireless networks are shown, these networks are arranged based on its operation range.

In Fig. 5.5, the different environment categories appear. These categories can be monitored by the WSN with utilizing the suitable scenarios [10, 11].

The familiar two technologies are suggested by many researches to implement the WSN application is the low range wireless technologies which are the ZigBee

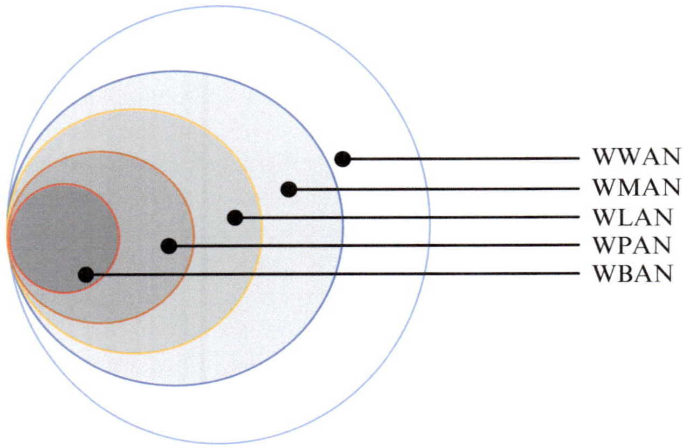

Fig. 5.4 The different wireless technologies

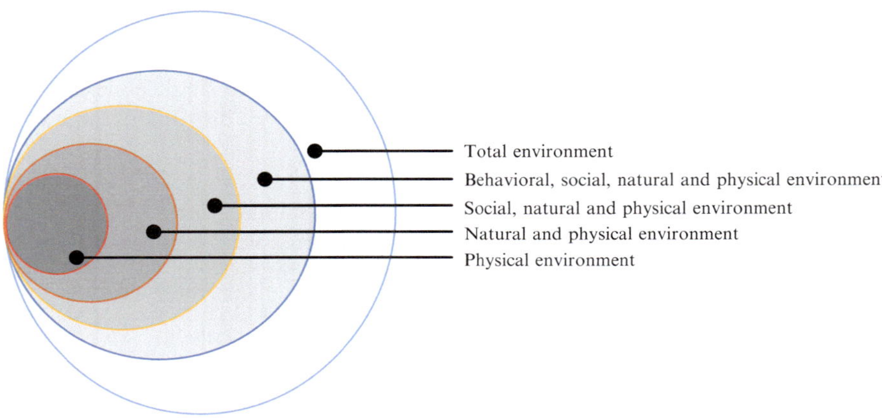

Fig. 5.5 The environment term contents

and Bluetooth networks. So, these networks can be used to implement the WSN application scenarios within the local application and limited service area. Out the range of these technologies, the application scenarios are implemented by executing the cooperation with higher ranges technologies. So, the WSN applications are not restricted by specific range.

5.5.1 IEEE 802.15.4 and ZigBee

ZigBee/IEEE 802.15.4 [3] is a bidirectional wireless technology featured with short range, low cost, low power consumption, and low data rate, which makes it more

Table 5.1 IEEE 802.15.4 frequency bands and data rates

PHY (MHz)	Freq band (MHz)	Mod.	Channels	Bit rate (kbps)
868/915	868–868.6	BPSK	1	20
	902–928	BPSK	10	40
2,450	2,400–2,483.5	O-QPSK	16	250

suitable for applications associated with monitoring and remote control that are integrated with functional sensors and actuators. Normally, ZigBee works in the license-free 2.4 GHz ISM band with a data rate of up to 250 Kbps and the transmission distance ranges from 10 to 75 m, depending on the power output and environmental dynamics [12].

5.5.1.1 Overview ZigBee Network

The ZigBee network is a Low-Rate WPAN (LR-WPAN) that is used for short-range and low-cost data communication. Table 5.1 gives brief of some of ZigBee technology specifications.

Low power consumption in ZigBee networks can be achieved by allowing a device to sleep, which means waking into active mode for brief periods. Enabling such low duty cycle operation is at the heart of the ZigBee standard [13]. ZigBee is built on top of the IEEE 802.15.4 standard. It offers the additional functionality to implement mesh networking rather than point-to-point networking found in most Bluetooth and Wi-Fi applications. The ZigBee specification document is short, allowing a small and simple stack, in contrast to the other wireless standards such as Bluetooth [14].

The IEEE 802.15.4 standard is intended to conform to established regulations in Europe, Japan, Canada, and the United States. It defines two physical (PHY) layers; the 2.4 GHz and 868/915 MHz band PHY layers. Although the PHY layer chosen depends on local regulations and user preference, only the higher data rate, worldwide, unlicensed 2.4 GHz Industrial, Scientific and Medical frequency band is considered [15]. A total of 16 channels are available in the 2.4 GHz band, numbered from 11 to 26, each with a bandwidth of 2 MHz, and a channel separation of 5 MHz. The channel mapping frequencies are given in Table 5.1. The LR-WPAN output powers are around 0 dBm. This wireless network typically operates within a 50-m range. The transmit scheme used is the Direct Sequence Spread Spectrum (DSSS) [16].

The ZigBee network involves little or no infrastructure. It also has a primitive error control mechanism, which is the Automatic Repeat reQuest (ARQ). As a result, this mechanism is unable to reduce the channel effects. So, there is a need for either a coding or interleaving mechanism to combat the bad channel effects [17]. The main advantages of the ZigBee technology are summarized as follows [7]:

1. Low power consumption.
2. Low data rate (250 Kbps).
3. Short distance.
4. Low cost, resulting from the low-rate and the simplicity of the protocol stack.
5. Possibility of up to 65,535 devices per network.
6. Robustness and reliable data transfer.

Several researchers have studied the transmission of images with the IEEE 802.15.4 standard. In [18], the authors studied the process of image fragmentation for transmission over the ZigBee network. In the case of transmission over mobile networks, there is a probability of burst errors. The burst errors have a bad effect on the transmitted data and image.

5.5.1.2 ZigBee Security

The IEEE 802.15.4 standard and the ZigBee specification are popular protocol choices for WSNs. Therefore, this chapter concludes with a discussion of the security measures available in these protocols.

The IEEE 802.15.4 standard provides four basic security models: *access control, message integrity, message confidentiality*, and *replay protection* [15]. Security in IEEE 802.15.4 is handled by the MAC layer and an application can choose specific security requirements by setting appropriate parameters in the radio stack (by default, security is not enabled). The standard distinguishes between eight security suites as shown in Table 5.2, each with different levels of protection for the transmitted data. The first suite offers no security, the second suite offers encryption only (AES – CTR), followed by a group of suites with authentication only (AES – CBC – MAC), and a group of suites with both authentication and encryption (AES – CCM). Suites that offer authentication differ in the sizes of the MAC, which varies from 32 to 128 bits. For every suite that offers encryption, IEEE 802.15.4 also offers optional replay protection consisting of monotonically increasing sequence numbers for messages to allow a recipient to detect replay attacks

The first suite Null does not provide any security. All other security suites use the Advanced Encryption Standard (AES) block cipher, which is also known as Rijndael. The National Institute of Standards and Technology defines five modes of operation, including the counter (CTR) and cipher block chaining (CBC) modes [15]. When authentication is needed, one of the three AES – CBC – MAC variants can be used, which compute a message integrity code using a block cipher in CBC mode. The three AES – CCM suites combine encryption and authentication by using the counter mode and the CBC mode (CCM is short for Counter with CBC – MAC).

In addition to the security features of IEEE 802.15.4, the ZigBee specification also introduces the concept of a trust center, a responsibility typically assumed by the ZigBee coordinator. The trust center is responsible for authentication of devices wishing to join a network (trust manager), maintaining and distributing keys

Table 5.2 Security suites supported in IEEE 802.15.4

Name	Description
Null	No security
AES-CTR	Encryption only, CTR mode
AES-CBC-MAC-128	128-bit MAC
AES-CBC-MAC-64	64-bit MAC
AES-CBC-MAC-32	32-bit MAC
AES-CCM-128	Encryption and 128-bit MAC
AES-CCM-64	Encryption and 64-bit MAC
AES-CCM-32	Encryption and 32-bit MAC

(network manager), and enabling end-to- end security between devices (configuration manager).

ZigBee also differentiates between a *residential* and a *commercial* mode. In the residential mode, the trust center allows nodes to join the network, but it does not establish keys with the network devices. In the commercial mode, it generates and maintains keys and freshness counters with every device in the network. The disadvantage of the commercial mode is its memory cost, which grows with the size of the network.

The ZigBee specification uses the CCM* mode for its security services, which is also a combination of CTR mode and CBC–MAC mode. Compared to the CCM mode, CCM offers encryption-only and integrity-only capabilities. Similar to the specifications in the IEEE 802.15.4 standard, ZigBee has several levels of security, including no security, encryption only, authenticated only, and both encryption and authentication. Levels that provide authentication use a MAC that can vary from 4 to 16 bytes.

5.5.2 Bluetooth Technology

Bluetooth technology was originally conceived by Ericsson in 1994. Ericsson, IBM, Intel, Nokia, and Toshiba formed the Bluetooth Special Interest Group (SIG) to drive development of Bluetooth products and serve as the governing body for Bluetooth specifications. Bluetooth is standardized within the IEEE 802.15 Working Group for Wireless Personal Area Networks that formed in early 1999 as IEEE 802.15.1-2002. The brief about the basic and EDR packets is presented in the previous Chap. 3, Sect. 3.3.

Word Bluetooth is a historical word. This name was intended to be temporary name but because no one has introduced a better name than it. The name follows a Danish king called Harald Blatand also known as Harald Bluetooth [15]. This king is famous for his success in uniting the Scandinavian people in the tenth century AD. Also, the Bluetooth standard is supposed to unite personal devices. Today, Bluetooth wireless technology enables electronic devices to talk to each other, but this time by means of a low-cost, short-range radio link.

5.5.2.1 Bluetooth Overview

Bluetooth is an open standard for short-range radio frequency (RF) communication. Bluetooth technology is used primarily to establish wireless personal area networks (WPAN), commonly referred to as ad hoc or peer-to-peer (P2P) networks. Bluetooth technology has been integrated into many types of business and consumer devices, including cell phones, laptops, automobiles, printers, keyboards, mice, and headsets. This allows users to form ad hoc networks between a wide variety of devices to transfer voice and data. This document provides an overview of Bluetooth technology and discusses related security concerns.

Several Bluetooth versions are currently in use in commercial devices. At the time of writing, Bluetooth 1.2 (adopted November 2003) and 2.0 + Enhanced Data Rate (EDR, adopted November 2004) are the most prevalent. Bluetooth 2.1 + EDR (adopted July 2007), which is quickly becoming the standard, provides significant security improvements for cryptographic key establishment in the form of Secure Simple Pairing (SSP). The most recent versions include Bluetooth 3.0 + High Speed (HS, adopted April 2009), which provides significant data rate improvements, and Bluetooth 4.0 Low Energy (LE, adopted June 2010), which supports smaller, resource-constrained devices and associated applications. This publication addresses the security of all these versions of Bluetooth.

Bluetooth technology and associated devices are susceptible to general wireless networking threats, such as denial of service (DoS) attacks, eavesdropping, man-in-the-middle (MITM) attacks, message modification, and resource misappropriation. They are also threatened by more specific Bluetooth-related attacks that target known vulnerabilities in Bluetooth implementations and specifications. Attacks against improperly secured Bluetooth implementations can provide attackers with unauthorized access to sensitive information and unauthorized use of Bluetooth devices and other systems or networks to which the devices are connected.

The Bluetooth technology is the most WPANs network type spreading. It can be considered the real engine of the personal communication developing. It rounded no-specialists people. This WPAN has many versions.

Figure 5.6 shows the original application of Bluetooth technology. These applications seem very simple, in the same time; it lets the wireless technology development very closed to the non-specialist peoples. It opens the gate of personal communication developing.

It can be said that Bluetooth technology has rounded people non-specialists from feeling including progress reached by the science and technology of digital communications and wireless networks. It also allowed the researchers to the idea of the introduction of wireless communication network technology in the areas were not in the mind or is expected to involve this kind of networks in such applications.

The Bluetooth technology proves a local area networking using ad hoc piconet cellular structure for wireless communications. It is an infrastructure less arrangement for communication within the local area networking. Through an access point infrastructure, this LAN can communicate with other wired and wireless LANs.

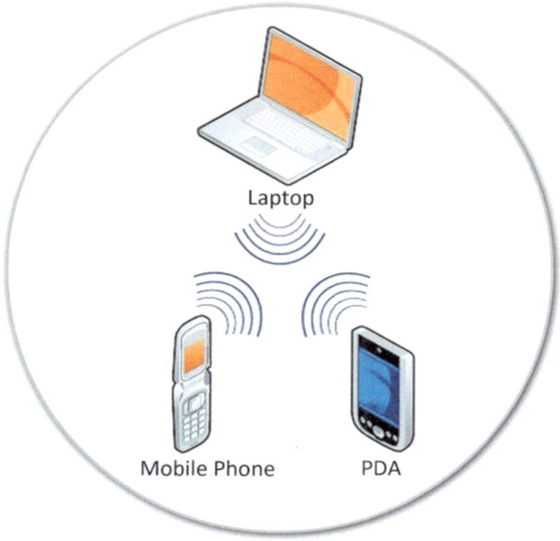

Fig. 5.6 Bluetooth ad hoc topology

The piconet ad hoc network arrangement will be quite useful in connecting a number of devices within an area without wires.

As shown in Fig. 5.7, there are three piconets Bluetooth network, it conforms scatternet. The presented scenario of Bluetooth devices connections and the merging between them is shown in Fig. 5.7 by using the same principle, the WSN can be implemented be the Bluetooth network also, the scenario overcomes the challenge of less active Bluetooth devices.

5.5.2.2 Versions Characteristics

Bluetooth devices can support multiple data rates using native Bluetooth and alternate Medium Access Controls (MAC) and Physical Layers (PHY). Because Bluetooth specifications are designed to be backward-compatible, a later specification device that supports higher data rates also supports the lower data rates supported by earlier specification devices (e.g., an EDR device also supports rates specified for BR devices). The following sections provide an overview for Bluetooth and alternate MAC/PHYs, as well as associated data rates and modulation schemes.

a. Basic Rate/Enhanced Data Rate

Bluetooth versions 1.1 and 1.2 only support transmission speeds of up to 1 megabit per second (Mbps), which is known as Basic Rate (BR), and can achieve throughput of approximately 720 kilobits per second (kbps). Introduced in Bluetooth version 2.0, Enhanced Data Rate (EDR) specifies data rates up to 3 Mbps and throughput of approximately 2.1 Mbps. BR uses Gaussian Frequency-Shift Keying (GFSK) modulation to achieve a 1 Mbps data rate. EDR

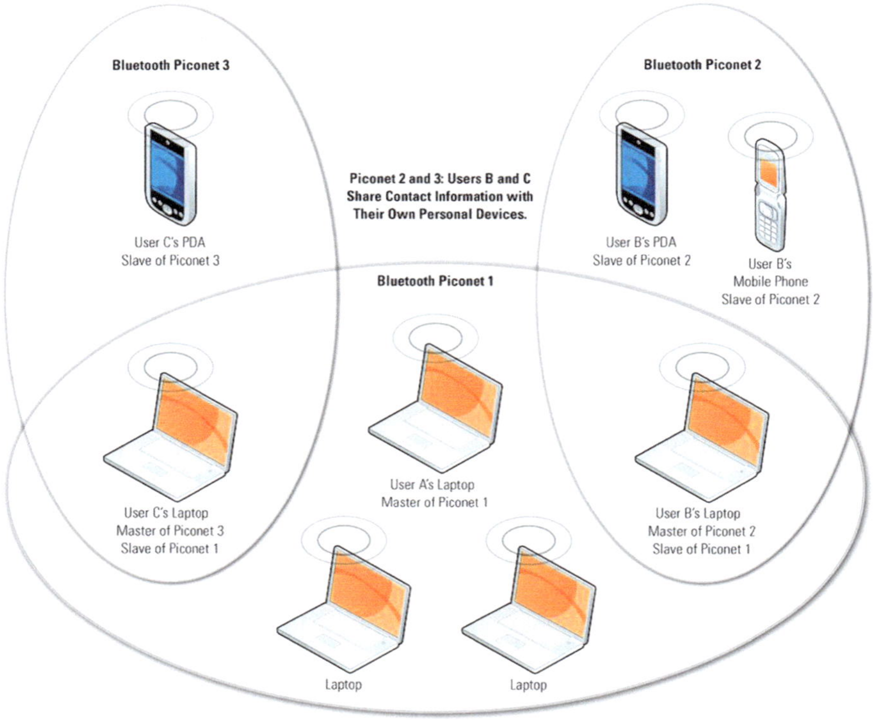

Fig. 5.7 Bluetooth networks (multiple piconets)

uses $\pi/4$ rotated Differential Quaternary Phase Shift Keying (DQPSK) modulation to achieve a 2 Mbps data rate, and 8 phase Differential Phase Shift Keying (8DPSK) to achieve a 3 Mbps data rate. Note that EDR support is not required for devices compliant with the Bluetooth 2.0 specification or later. Therefore, there are devices on the market that are "Bluetooth 2.0 compliant" versus "Bluetooth 2.0 + EDR compliant." The former are devices that support required version 2.0 features but only provide the BR data rate.

b. High Speed with Alternate MAC/PHY

Introduced in the Bluetooth 3.0 + HS specification, devices can support faster data rates by using Alternate MAC/PHYs (AMP). This is known as Bluetooth High Speed (HS). In the Bluetooth 3.0 + HS specification, IEEE 802.11-2007 was introduced as the first supported AMP. IEEE 802.11-2007 is a rollup of the amendments IEEE 802.11a through 802.11j. For the 802.11 AMP, IEEE 802.11 g PHY support is mandatory, while IEEE 802.11a PHY support is optional. The 802.11 AMP is designed to provide data rates up to 24 Mbps using Orthogonal Frequency-Division Multiplexing (OFDM) modulation. Note that this AMP is IEEE 802.11 compliant

Table 5.3 Key differences between Bluetooth BR/EDR and LE

Characteristic	Bluetooth Br/EDR	Bluetooth LE
RF physical channel	79 channels with 1 MHz channel spacing	40 channels with 2 MHz channel spacing
Discovery/connect	Inquiry/paging	Advertising
Number of piconet slaves	7 active/255 total	Unlimited
Device address privacy	None	Private device addressing available
Max data rate	1–3 Mbps	1 Mbps via GFSK modulation
Encryption algorithm	E0/Safer+	AES-CCM
Typical range	10–100 m	50 m
Max output power	100 mW (20 dBm)	10 mW (10 dBm)

but not Wi-Fi compliant. Therefore, Wi-Fi Alliance specification compliance is not required for Bluetooth 3.0 + HS devices.

c. Low Energy

Bluetooth LE was introduced in the Bluetooth 4.0 specification. Formerly known as "Wibree" and "Ultra Low Power Bluetooth," LE is primarily designed to bring Bluetooth technology to coin cell battery powered devices such as medical devices and other sensors. The key technology goals of Bluetooth LE (compared with Bluetooth BR/EDR) include lower power consumption, reduced memory requirements, efficient discovery and connection procedures, short packet lengths, and simple protocols and services. Table 5.3 provides the key technical differences between BR/EDR and LE.

5.5.2.3 Security Features

Bluetooth has three different modes of security. Each Bluetooth device can operate in one mode only at a time. These three modes are security mode (1) non-secure mode, security mode (2) service-level enforced security mode, and security mode (3) link-level enforced security mode.

In security mode 1, there is no security procedure. In this mode, the security functionality such as authentication and encryption is bypassed. In security mode 2, the security procedures are initiated after the connection establishment at Logical Link Control and Adaptation Protocol (L2CAP). In security mode 3, which is the link-level security mode, a Bluetooth device initiates the security procedure before the channel is established [12].

Cumulatively, the family of Bluetooth BR/EDR/HS specifications defines four security modes. Each Bluetooth device must operate in one of these modes, called Security Modes 1 through 4. These modes dictate when a Bluetooth device initiates security, not whether it supports security features. Security Mode 1 devices are considered non-secure. Security functionality (authentication and encryption) is never initiated, leaving the device and connections susceptible to attackers. In effect, Bluetooth devices in this mode are "indiscriminate" and do not employ any mechanisms to prevent other Bluetooth-enabled devices from establishing connections. However, if a remote device initiates security such as a pairing, authentication, or encryption request a Security Mode 1 device will participate. Per their respective Bluetooth specification versions, all v2.0 and earlier devices can support Security Mode 1, and v2.1 and later devices can use Security Mode 1 for backward compatibility with older devices. However, NIST recommends never using Security Mode 1 [6].

In Security Mode 2, a service level-enforced security mode, security procedures may be initiated after link establishment but before logical channel establishment. For this security mode, a local security manager (as specified in the Bluetooth architecture) controls access to specific services. The centralized security manager maintains policies for access control and interfaces with other protocols and device users. Varying security policies and trust levels to restrict access can be defined for applications with different security requirements operating in parallel. It is possible to grant access to some services without providing access to other services. In this mode, the notion of authorization the process of deciding whether a specific device is allowed to have access to a specific service is introduced. It is important to note that the authentication and encryption mechanisms used for Security Mode 2 are implemented in the controller, as with Security Mode 3 described below. All v2.0 and earlier devices can support Security Mode 2, but v2.1 and later devices can only support it for backward compatibility with v2.0 or earlier devices [7].

Security Mode 3 is the link level-enforced security mode, in which a Bluetooth device initiates security procedures before the physical link is fully established. Bluetooth devices operating in Security Mode 3 mandate authentication and encryption for all connections to and from the device. All v2.0 and earlier devices can support Security Mode 3, but v2.1 devices can only support it for backward compatibility purposes [2].

5.6 Summary

Like every other computer network, wireless sensor networks are exposed to a variety of threats and attacks and like most other networks, sensor networks require support for confidentiality, integrity, and authentication to protect sensor nodes and sensor data. However, several unique characteristics of WSNs, such as remote deployment (which facilitates an adversary's physical access to sensor nodes) and resource constraints make it easier to compromise sensors and sensor data. Further, many sensor networks are attractive targets for attackers due to the nature of many WSN applications and the sensitive data they generate (e.g., military applications, emergency response, health care). This chapter provided a brief overview of several types of attacks commonly found in sensor networks and techniques and protocols to defend a network or to detect an intrusion or compromised node. As WSNs continue to become more commonplace, it is to expect that security challenges will increase, the types and number of threats will evolve, and new solutions to protect sensor networks and sensor data will be required.

References

1. Roy DB, Chaki R, Chaki N (2009) A new cluster-based wormhole intrusion detection algorithm for mobile ad-hoc networks. Int J Netw Secur Appl (IJNSA) 1(1): 44–52
2. Shanti N, Ganesan L, Ramar K (2009) Study of different attacks on multicast mobile ad-hoc network. J Theor Appl Inform Technol 45–51
3. Wei C, Xiang L, Yuebin B, Xiaopeng G (2007) A new solution for resisting gray hole attack in mobile ad-hoc networks. In: Second international conference on communications and networking in China, pp 366–370
4. Parsons M, Ebinger P. Performance evaluation of the impact of attacks on mobile ad-hoc networks. https://www.cse.buffalo.edu/duvms2009_submission_parsons.pdf
5. Bloom JA, Cox IJ, Kalker T, Linnartz JPMG, Miller ML, Traw CBS (1999) Copy protection for DVD video. Proc IEEE 87:1267–1276
6. Chen M, He Y, Lagendijk RL (2005) A fragile watermark error detection scheme for wireless video communications. IEEE Trans Multimed 7:315–329
7. Luis P-F, Pedro C, Juan RT-P, Fernando P-F (2006) Watermarking security: a survey. LNCS Trans Data Hiding Multimed Secur 1:41–72
8. Chuhong F, Kundur D, Kwong RH (2006) Analysis and design of secure watermark-based authentication systems. IEEE Trans Inf Forensics Secur 1:43–55
9. Jani PV (2002) Security within ad-hoc networks. Position paper, PAMPAS workshop, 16/17 Sept 2002
10. Pathan ASK, Hyung L, Hong CS (2006) Security in wireless sensor networks: issues and challenges. In: Proceedings of 8th IEEE international conference on advanced communication technology, Phoenix Park, vol 2, pp 1043–1048
11. Perrig A, Przydatek B, Song D (2007) SIA: Secure Information Aggregation in sensor networks. J Comput Secur 15:69–102
12. Chen M, He Y, Lagendijk RL (2001) Error detection by fragile watermarking. In: Proceedings of the 22nd of the picture coding symposium, Seoul, pp 287–290
13. Ren X, Yu H (2006) Security mechanisms for wireless sensor networks. Int J Comput Sci Netw Secur 6:151–161
14. Dargie W, Poellabauer C (2010) Fundamental of wireless sensor networks theory and practice. Wiley, Series on Wireless Communications and Mobile Computing
15. Sastry N, Wagner D (2004) Security considerations for IEEE 802.15.4 networks. In: Proceedings of the 3rd ACM workshop on wireless security, Philadelphia
16. Durresi A, Paruchuri V, Kannan R, Lyengar SS (2005) Data integrity protocol for sensor networks. Int J Distrib Sens Netw 1:205–214
17. Venugopalan R, Ganesan P, Peddabachagari P, Dean A, Mueller F, Sichitiu M (2003) Encryption overhead in embedded systems and sensor network nodes: modeling and analysis. In: Proceedings of international conference on compilers, architectures and synthesis for embedded systems, San Jose, pp 188–197. Sensors 2011, 11:4135
18. Hartung F, Kutter M (1999) Multimedia watermarking techniques. Proc IEEE 87:1079–1107

Chapter 6
Lower Complexity of Secured WSN Networks

Abstract In this chapter, the WSN performance is investigated based on the most common WPANs technologies, the ZigBee and Bluetooth networks. Supper interleaving versions have been presented based on the encryption tools with an additional complexity is considered. These techniques are employed to produce performance enhancing with keeping the security capability. It discussed the traditional interleaving techniques for data and image transmission over fixed and mobile WSN. The mobility of WSN nodes challenges and feature also, are presented.

Keywords Newest interleaving techniques • Mobile WSN • FEC complexity • EMSN

The mobility of WSN is discussed, and its effects and its need also. There is a new version of interleaving techniques based on the chaotic idea is devoted in this chapter. The simulation experiments computer based are presented to investigate the mobility effects also, to investigate the different scenarios. The historical brief of the interleaving chaotic based technique is discussed. The last point in the chapter, the proposed cooperative WSN algorithm is presented with different scenarios.

The chapter also discusses the complexity of error control schemes. It is added to the total system complexity. Almost, the powerful technique has higher computational complexity. The complex FEC will not be needed in the presented scenarios by applying powerful and new versions of interleaving techniques. So, the complexity for all system will be decreased.

6.1 Power with FEC

The chapter also discusses the complexity of error control schemes. It is added to the total system complexity. Almost, the powerful technique has higher computational complexity. The error control schemes are utilized in the wireless communication to achieve decreasing the power needed to reach specific (Bit Error Rate) BER, or to increase the throughput of the system, or reducing the time of

M.A.M. El-Bendary, *Developing Security Tools of WSN and WBAN Networks Applications*, Lecture Notes in Electrical Engineering 316, DOI 10.1007/978-4-431-55069-3_6

transmission by decreasing the retransmission times. Minimizing transmit power is considered herein to be the most critical parameter for a low-power WSN, whose battery lifetime is dependent on power consumption. Therefore all transmit power and energy calculations use the minimum required transmit power and energy. In a low-power WSN scenario, transmitting with as much power as possible, up to regulatory limits, is not desirable. Rather, transmitting with as little power as possible, so as to extend sensor battery life, while maintaining a minimum required SNR, is the main target in this work scenarios, the low-power WSN is far more power-constrained than bandwidth-constrained. In order to achieve power efficiency, we are willing to sacrifice spectral efficiency.

The complexity of the system can be defined as the number of all arithmetic and logic operations are carried out through the system and the memory length is need for this operations. So, length of the processed data can be one of the factors which affects on the system complexity. Lower complexity means lower operations, lower time of processing, lower delay, and lower power consumption, and so on,

Fist, let us see the role of FEC effect on the transmitted power. The Signal to Noise Ration (SNR or S/N) is defined as the minimum required power of the wireless link to be reliable, lower than this value link will be unreliable.

So, as shown in the following equation [1],

$$\frac{S}{N} = \frac{R}{B} \frac{E_b}{N_0} \tag{6.1}$$

R is the information rate.
B is the signal bandwidth.

The ratio of the information rate R to the bandwidth is defined as the spectral efficiency (η).

The signal of noise can be expressed by the following:

$$N = mkTB \tag{6.2}$$

m is the noise proportional constant.
k is the Boltzmann constant.
T is the absolute temperature in K.

As shown the Eq. 6.2, the noise is proportional to the thermal noise and the bandwidth.

The receiver noise figure (RNF) in dB is incorporated in to the proportional constant (m) $m > 1$. In case of ideal receiver $m = 1$.

$$m = 10^{RNF/10}$$

To find the relation between the received and transmitted power, let us suppose the following:

The received power is SRx = S, at the d distance, where d is the distance separation between the transmitter and receiver in meter.

So, the received power can be expressed by the formula Eq. 6.3 with assuming an omnidirectional antenna type and no interference no obstacles.

$$S_{Rx} = \left(\frac{1}{4\pi d^2}\right)\frac{\lambda^2}{4\pi}P_{Tx} \tag{6.3}$$

The right side of previous Eq. 6.3 is related to the distance effect, which is denoted by γ.

$$\gamma = \left(\frac{1}{4\pi d^2}\right)\frac{\lambda^2}{4\pi} = \left(\frac{\lambda}{4\pi d}\right)^2 \tag{6.4}$$

The λ is wavelength of transmitted signal, $\lambda = \frac{C}{F}$, C is the light speed $C = 3\times10^8 m/\mathrm{sec}$ and F refers to the transmitted signal frequency. So, the transmitted power can be formulated from Eq. 6.3 as follows in Eq. 6.5 [2]:

$$P_{Tx} = \left(\frac{4\pi d}{\lambda}\right)^2 P_{Rx} \tag{6.5}$$

The received power is the $P_{Rx} = S$, inserting the noise by multiplying the right side by *N*/*N* and compensating in the Eq. 6.6.

$$P_{Tx} = \frac{S}{N}N\left(\frac{4\pi d}{\lambda}\right)^2 \tag{6.6}$$

Using the Eq. 6.1, the transmitted power will be in the following form:

$$\begin{aligned} P_{Tx} &= \frac{R}{B}\frac{E_b}{N_0}mkTB\left(\frac{4\pi d}{\lambda}\right)^2 \\ &= \gamma\frac{E_b}{N_0}mkTB\left(\frac{4\pi d}{\lambda}\right)^2 \end{aligned} \tag{6.7}$$

From the Eq. 6.7

- The transmitted power proportional to the square distance which the transmitted signal needed to travel or cross.
- Also, it inversely proportional to the square transmitted signal wavelength. So, lower wavelength means higher transmitted power. That means also, the transmitted power proportional to the transmitted signal frequency. There also, many factors affects on the transmitted power such as the distance, the path loss,

reduction of Line-of-Sight (LOS) components, interference and the attenuation [3].

The path loss can be expressed in the following equation:

$$\text{path Loss} = PL(d) = \frac{P_{Rx}}{P_{Tx}} = \left(\tfrac{4\pi d}{\lambda}\right)^2 \text{ in Log scale the equation will be : -} \\ = PL(d_0) + 10n\log_{10}\left(\frac{d}{d_0}\right) \tag{6.8}$$

In the path loss equation, n is the path loss index (n = 2 for free space (LOS) and lies between 2 and 4 for non LOS), d0 refers to the standard distance between the transmitter and the receiver antenna d0 = 1 m, and the d is the distance of the signal travels in general.

So, the path loss due to the d0 distance can be expressed by:

$$PL(d_0) = 20\log_{10}\left(\frac{4\pi d_0}{\lambda}\right) \tag{6.9}$$

In brief, the transmitted power is a function of the following variables:

- The distance.
- The path loss index.
- The receiver's antenna gain and effective area, which are ignored in the previous analysis.

A line-of-sight (LOS) communication link, the path loss index is 2. A non-LOS communication link, it lies between 2 and 4.

In the presented scenarios in the following section, the coding gain of weak error control scheme enhanced and improved with notified degree. It performs better than the more complex error control schemes. The presented scenario are carried out and investigated for evaluate the proposed technique which is called "Chaotic Interleaved Error Control Schemes" (CIECS). This technique utilizes the block code type such as the Hamming (15,10) code and the convolutional codes (1,2) with different constraint lengths. The data randomizing tool here depends on the secret key in its job, so, the presented scenario provides security addition over the powerful performance. It reduced the need of complex encoder and decreases the required SNR.

In the following, the relation between the transmitted consumption saving and the channel coding utilization is presented and analyzed for the WSN. Almost of WSN application needs long life batteries, many application field in unreachable places or nor easy reachable locations. The WSN utilization principle is "***operate me and run***" or "***operate me and leave me sense***"

So, the transmitted power as shown in the previous Eq. 6.7 can be rewritten as follows [4]:

$$P_{Tx} = \gamma \frac{E_b}{N_0} N \left(\frac{4\pi d}{\lambda}\right)^2 = \gamma 10^{(SNR/10)} 10^{RNF/10} (kTB) \left(\frac{4\pi d}{\lambda}\right)^2 \tag{6.10}$$

The difference between the uncoded signal and encoded one treatment is the SNR should be changed and the spectral efficiency γ, where the transmission information rate is changed also. In the following Eq. 6.11, it gives the transmitted power of uncoded system formula [5].

$$P_{U.Tx} = \gamma_u 10^{(SNR_u/10)} 10^{RNF/10} (kTB) \left(\frac{4\pi d}{\lambda}\right)^2 \tag{6.11}$$

U in this equation denoted to no FEC, the signal is transmitted without EER protection. The SNR_u as mentioned in previous is the required power to noise ratio to achieve specific BER performance for the uncoded transmitted signal. In the simulation Computer experiments, the Binary Phase Shift Keying (BPSK) modulation is utilized, so, the spectral efficiency of the BPSK for uncoded signal is R/B = 1.

In case of coded signal transmission, there are extra bits. So, the number of transmitted bits increases, which leads to the information transmission rate (R) will be decreased. Let us suppose the code rate (Rc). The transmission rate of encoded signal (RE) is given in Eq. 6.12 [6]:

$$R_E = R_C \times R \tag{6.12}$$

So, the transmitted power of the encoded signal can be expressed by:

$$P_{E.Tx} = \gamma_E 10^{(SNR_E/10)} 10^{RNF/10} (kTB) \left(\frac{4\pi d}{\lambda}\right)^2 \tag{6.13}$$

In the last, it can be briefed in two expressions:

$\gamma_E < \gamma_U$	FEC occupy part in the channel capacity
$SNR_E > SNR_U$	FEC enhance the power efficiency

The transmission rate of uncoded system is higher than the encoded system. The encoded system really decreases the minimum required power also, it decreases the number or dropped packets. So, it decreases the need to more retransmission times. The channel conditions determine the value of channel coding utilization.

There are two factors let us propose the FEC schemes for the WSN mechanism, first, the FEC decreases the needed power which is required to achieve the same BER. The power factor is the main challenges in the WSN applications.

The second, the most of WSN applications is not need high data rate transmission, it need much more in desperate need of reliable link and lower delay.

6.2 Complexity FEC Dependant

The transmission of multimedia contents over unreliable data links has become of paramount importance. This type of transmission must reconcile the high data rates involved in multimedia contents and the noisy nature of the channels, be it wireless or mobile. In our simulation, we try to improve the transmission of colored image over Bluetooth network through different interleaving and error control schemes. Also we discuss the fragmentation process in different cases. We use an interleaving technique before transmission process. We use block interleaver, Convolutional codes, standard error control Bluetooth scheme, and chaotic interleaving (proposed).

6.2.1 Error Control Basics

The utilization of WSN fields is related to the environmental monitoring and open areas, so, the data of captured events are transmitted over a noisy channel. The reliability of wireless link for the WSN applications is essential point. The employing of FEC schemes for WSN packets encoding is granted. In this section, the important terms of error control techniques are presented. The extra computational complexity of the WSN system due to the error control techniques utilization is analyzed also.

The purpose of the use of different error control schemes represents in two goals, error detection and error correction, or both together.

There are many differences between block codes and convolutional codes [7]. These two channel codes are used for detection and correction channel bit errors. This chapter presents a block codes important terms in general. Where block codes accept a block of k information bits as input encoder and produce a block of n information bits as output of encoder [8]. There are many types of block codes are commonly used, these block codes as Hamming codes, Golay codes, and BCH codes. The channel code, which is used for Bluetooth system is block codes, BCH code, repetition code, and Hamming code these block codes are used in Bluetooth frame.

There are important terms, which determine the rate of error control code and the number of redundant bits with the efficiency of error control codes.

Important Terms Definitions:

Code word (n): it is the output of encoder (n bits block), it is known as code word. It consists of the data word bits plus redundant bits.

$$n = k + m \tag{6.14}$$

Data word (k): it is the input of encoder, k bits block, it is known as data word. It is the difference between code word n and redundant bits m.

$$k = n - m \tag{6.15}$$

Code rate (R): it is the radio of the input of encoder (k), data word to the output of encoder (n) code word.

$$R = k/n \tag{6.16}$$

Code efficiency: this term is equal the code rate R.

$$\text{Code efficiency} = k/n$$

Error control codes have two methods to achieve its function:

1. Error detection with retransmission.
2. Error detection and correction.

In case of first type, after receiving the error will be detected, so the receiver will send Automatic Repeat reQuest (ARQ) to transmitter. This method cause a lot of time wasting. In case of second type the errors will be corrected after detected, hence the receiver will not send ARQ. The number of bits which are added by encoder = (bits at the output of encoder) – (bits at input of encoder) m, as shown in Fig. 6.1. The redundant bits which cause error detection or detection and correction of error control codes capabilities [9].

6.2.2 *Convolutional Codes*

There are some noted about the communication systems complexity of FEC dependent, it is briefed in the following: About the complexity of the channel coding difference between the convolutional codes and the block codes, it is restricted by the nature of every error control scheme type [10].

In Fig. 6.2, the construction of nonsystematic convolutional encoder is shown. The computational complexity of the convolutional codes encoding and decoding process is influenced by number of factors such as the number of input bits (k), and the output bits (n) also, the memory length of the encoder (m), as shown in Fig. 6.2, these factors are cleared.

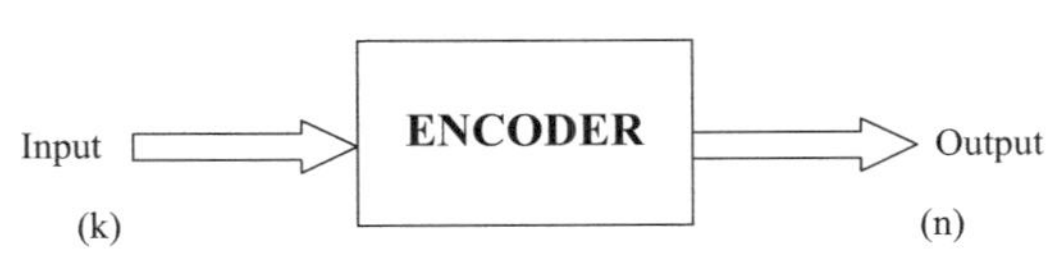

Fig. 6.1 An encoder block, its input, and its output

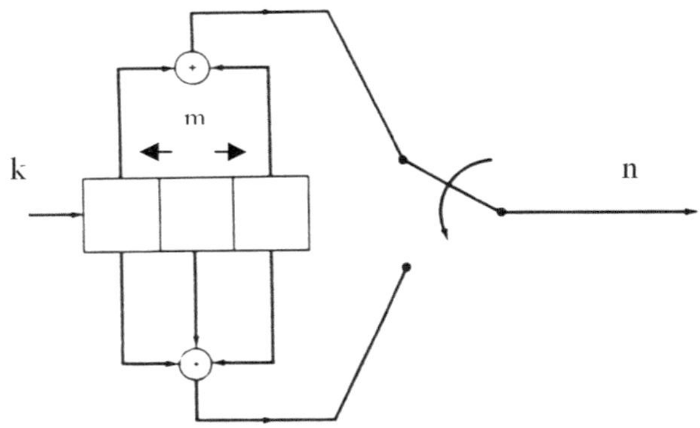

Fig. 6.2 Simple construction of the convolutional encoder

6.2.3 *Convolutional Code Gain*

The performance of convolutional codes is determined by its decoding algorithm and its distance properties. The minimum free distance d_{free} is a most important measure. It is defined as the minimum distance between two codewords in the code.

$$d_{free} = \min\{w(b)|a_1 \neq a_2\} \tag{6.17}$$

When a1 and a2 have different length, the shorter sequence is padded with zeros. The definition of the weight of a codeword is similar to that of block codes but for a codeword, it is produced by a nonzero input information sequence. As a linear code, we also have

$$d_{free} = \min\{w(b)|a_1 \neq 0\} \tag{6.18}$$

The d_{free} is the minimum weight of the codeword produced by any nonzero information sequence. For a given rate and encoder memory length, the free distance of a non-systematic code is larger than that of a systematic code.

Coding gain of the convolutional codes
The coding gain for the convolutional codes over an uncoded BPSK or QPSK system is upper bounded by [11].

$$G_C = 10\log_{10}\left(\frac{R_C d_{free}}{2}\right)\ dB \tag{6.19}$$

In the hard decision case the corresponding gain in the soft decision case is given by [11].

$$G_C = 10\log_{10}\left(R_C d_{free}\right)\ dB$$

So, the soft-decision introduces an additional 3 dB in the asymptotic coding gain.

Coding gain – the reduction in the required E_b/N_0 to achieve a specified BER of an error-correcting coded system

The coding gain is the amount of additional SNR or E_b/N_0 that would be required to provide the same BER performance for an uncoded signal

The minimum distance d_{free}, the corresponding GC, and the generators for a number of binary, short constraint length K convolutional codes at several code rate Rc including, Rc = 1/2, n = 2,. . .,8 and 3/4 have been tabulated in [12], the listed of d_{free} are the largest possible values for the given Rc and K, these codes are non-systematic codes, since for a given Rc and encoder memory, d_{free} is larger for non-systematic codes than the systematic codes. The value of the d_{free} can be increased by decreasing the code rate Rc or by increasing the K.

It is known that Convolutional codes achieve good performance by expanding their memory depth. Also, the complexity of the convolutional codes decoder Viterbi algorithm is increased exponentially with the memory depth. The coding gain of the convolutional codes is expressed by Eq. 6.20 as follows:

$$R_C = k/n \tag{6.20}$$

$$\begin{aligned} G_C &= kd_{free}/(2n) \\ &= R_C d_{free}/2 \end{aligned} \tag{6.21}$$

Where R_C is the code rate and the d_{free} is the free distance of the convolutional code.

Table 6.1 shows some of selected convolutional codes and their code gains $R_C d_{free}/2$ expressed for hard decoding also by Eq. 6.22:

$$G_{C\_dB} = 10\log_{10}\left(R_C d_{free}/2\right)\ dB \tag{6.22}$$

Table 6.1 Selected convolutional codes with their variables and its coding gain

n	k	R_c	L	d_f	Gain
4	1	1/4	3	13	1.63
3	1	1/3	3	10	1.68
2	1	1/2	3	6	1.50
			6	10	2.50
			9	12	3.00
3	2	2/3	3	7	2.33
4	3	3/4	3	8	3.00

6.2.4 Convolutional Code Complexity

As shown in Fig. 6.1, there are some factors determine the computational complexity of the convolutional codes as the following:

1. The convolutional codes are very common applied on the communications systems; it is not restricted by number of errors which can be corrected in the codeword. The utilization of these codes depends on it complexity factors.
2. The length of input encoder data (k) which is processed at a time.
3. The length data stream (L) is encoded as one data word; it is not divided to short length data word.
4. In block codes, the L is divided to data-word with limited k bits length.
5. In the convolutional codes, k refers to the number of input bits at one time interval, not the number of data word.
6. The decoding of block codes is not complex like the convolutional codes.
7. The Length of data increases the computational complexity of the convolutional encoder (2^k*L).

The computer simulations are based on the standard error control scheme of Bluetooth network which is Hamming code (15, 10) and binary non-recursive convolutional coding. The Convolutional codes are denoted by (n, k, m), where m is encoder Memory depth (number of register stages). The Constraint length $K = (m + 1)$ is defined as the number of encoded bits a message bit can influence to. The Convolutional codes achieve good performance by expanding their memory depth. The convolutional encoder uses a constituent encoder with $K = 3$, 7, or 9, code rate $= 1/2$ and generator polynomials, $G = (5, 7)$, (133,171), or (561,753) in octal [13]. The scenario uses the proposed chaotic interleaving technique to improve error performance the capabilities of the weak block code which is represented in this work by Hamming code (15, 10) compared to the performance of the convolutional codes with long constraint lengths to reduce the complexity of the communications system in general [14].

The block codes are defined by (n, k) where k is the number of input data bits and n is the number of bits in the encoded frame. These codes have low complexity than the other coding schemes. The computational complexity of the block codes is determined by the values of n and k. In the scenario, the error control scheme which is used for payload encoding is Hamming code (15, 10) [8]. It is error control scheme of old Bluetooth versions, $n = 15$ and $k = 10$. It has lower complexity where the computational complexity increases with high value of n [11]. According to the computational complexity of the most common convolutional codes decoder which is the Viterbi Algorithm (VA), it will be as follows:

$$O\left[\left(2^{k-Encoder.Input}\right)\left(2^{m-memory\ depth}\right)(L+m)\right] \tag{6.23}$$

According to Eq. 6.23 from [12] the general equation of the computational complexity of the convolutional codes decoding algorithm (VA) can be formulated as follows:

$$O\left[\left(2^{k=Encoder.input}\right)\left(2^{m=memory.depth}\right)(L+m)\right] \tag{6.24}$$

From this equation we can see it can be divided to sections as follows:

$$O\left[\left(2^{k=Encoder.input}\right)\left(2^{m=memory.depth}\right)L+\left(2^{k=Encoder.input}\right)\left(2^{m=memory.depth}\right)m\right] \tag{6.25}$$

First section due to the length of the encoder data input (L). The second section refers to the complexity increasing due to the memory depth of the convolutional encoder.

Notes: the second section may be neglected if the L is very longer than m (L >> m)

If the input of encoder is one bit at interval time so, the k = 1. Then the pervious equation will be as follows:

$$O\left[\left(2^{k=1}\right)\left(2^{m=memory\ depth}\right)L+\left(2^{k=1}\right)\left(2^{m=memory\ depth}\right)m\right] \tag{6.26}$$

Then the computational complexity of the VA of the convolutional encoder with one bit input and the memory depth is m can be written as follows:

$$O\left[\left(2^{m+1}\right)L+\left(2^{m+1}\right)m\right] \tag{6.27}$$

From this equation we observed that the m + 1 equal the constraint length of the convolutional codes (K) where K = m + 1, so, the equation is rewritten to be in the following form

$$O\left[\left(2^{K=Constraint\ Length}\right)(L)+\left(2^{K=Constraint\ Length}\right)m\right] \tag{6.28}$$

So, from the previous analysis it can be revealed that the extra computational complexity of the convolutional codes due to the memory depth of the convolutional encoder cannot be always neglected especially if the length of data input (L) was not very long proportion to the length of m [15–17].

The powerful of the presented scenarios of this work depends on the truth of the computational complexity of the block and convolutional codes. The computational complexity of the block codes depends on the length of k and n as mentioned previously, it doesn't depend on the length of data bit stream. If we suppose that L is the input data length, so, it divided to packets with k bits length. On the other hand, for the convolutional codes it depends on the L [18, 19].

6.3 Interleaving Techniques

In this section, the interleaving technique benefits are presented. The interleaving idea is utilized to limiting the need of complex error control schemes. The traditional interleaving techniques such as the block interleaving and the convolutional interleaving are compared to other interleaving techniques which are based on the encryption data tools. The last types presents powerful randomizing engine, furthermore, it enhances the wireless link security, where the randomization process is carried out using a secret key. This secret key can be unique for all transmitted data or changed every transmitted packet. In 2004, Zhang et al. employed the chaotic interleaver based on the logistic map as an interleaver scheme is utilized within the turbo code. So, in the following the definition and interleaver parameters are introduced. It will be followed by studying the different interleaving techniques using the same data stream.

A random-error-correcting code could perform much worse in a bursty channel than in a memoryless channel, even though the channel BER of the former is much lower than that of the latter. The idea is to use an interleaver to permute the channel input sequence before transmission, and to undo the permutation after transmission by a de-interleaver.

The error correction codes are effective techniques to combat the random errors. With the burst error or dependent error these techniques alone will be inefficient. The interleaving technique is art of change the original order and randomizing the data with specific manner using the device is called "*Interleaver*" [9]. As shown in Fig. 6.3.

Interleaver definition:
An (n1, n2) Interleaver is defined as a device that reorders a input sequence so that no continuous sequence of n2 symbols in the reordered sequence contains any symbols that are separated by fewer than n1 symbols in the original ordering. This device is employed within the communication systems in two forms:

1. Stand alone Interleaver (Pseudo-random coding scheme)
2. Interleaved Forward Error Correction.

Concatenated coding technique attempts to combine the performance-improvement of two or more codes. Concatenation of codes can follow one of three architectures, serial, parallel, or multilevel, each of which are introduced in the three following subsections.

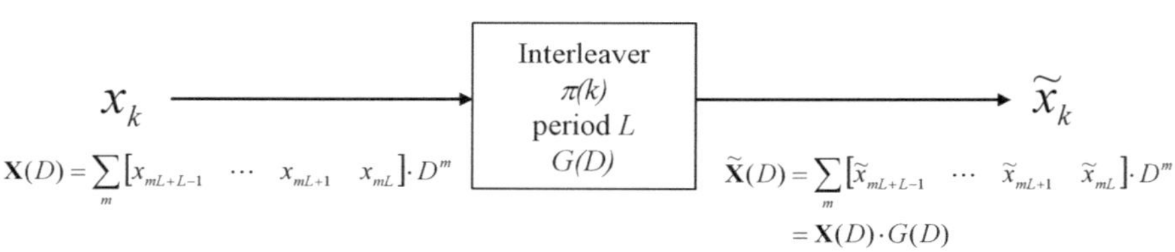

Fig. 6.3 Basic interleaver

6.3.1 *Interleaving Definition*

Interleaving is a periodic and reversible reordering of blocks of L transmitted symbols. Symbols (or bits) are correspondingly reordered by de-interleaving in the receiver. Interleaving is used to disperse error bursts that may occur because of non-stationary channel noise that may be localized to a few dimensions. Error bursts can also occur because of the incorrect decision of an inner/first decoder in Fig. 6.4 or 6.5. Error symbols caused by inner-code detection errors will typically span the entire incorrect codeword, leading to a burst of error symbols. If these bursts are separated by an interval long with respect to the interleaver period, then they can be distributed more evenly over time (or more generally dimensions) by the de-interleaver in the receiver. The distribution of the errors effectively enables realistic modeling for the inner-code-and-channel as memoryless, i.e., modeled by a Binary Symmetric Channel (BSC), Discrete Memory less Channel (DMC), or other channel for which successive symbol outputs are independent. Figure 6.4 generically depicts the interleaver as accepting symbols indexed in time by k or in block/packet in m, where L symbols occur within one packet and L is the period of the interleaver. Often L corresponds to a codeword size.

Definition 1 (Depth of an Interleaver) *The* **depth** J *of an interleaver is defined as the minimum separation in symbol periods at the output of the interleaver between any two symbols that were adjacent at the input of the interleaver.*

The depth of an interleaver has significant implication for a burst of errors entering a de-interleaver at a receiver. If a burst of errors has duration less than the depth, then two symbols affected by the burst cannot be adjacent after de-interleaving.

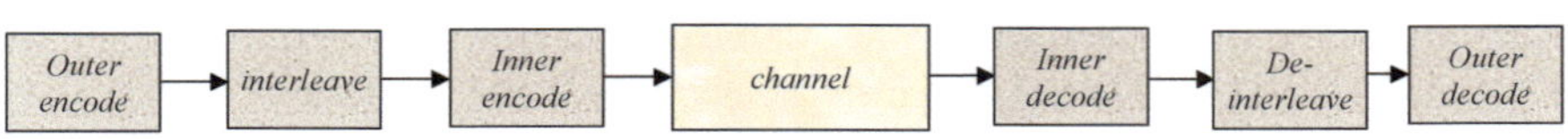

Fig. 6.4 Serial concatenation

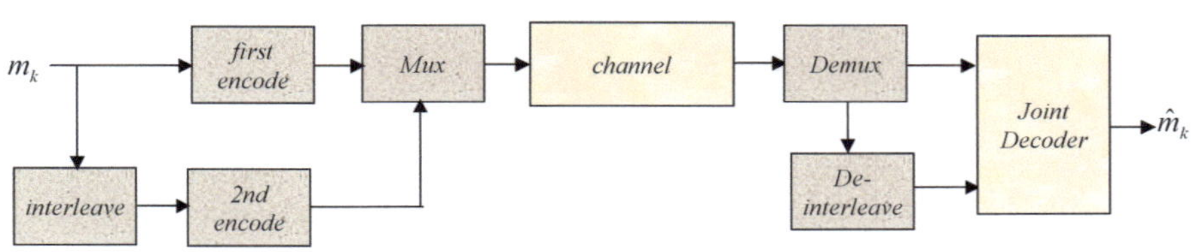

Fig. 6.5 Parallel concatenation

Definition 2 (Period of an Interleaver) *The* **period** L *of an interleaver is the shortest time interval for which the re-ordering algorithm used by the interleaver repeats.*

Essentially, the period is established by the detailed description of the interleaver and measures the length of a block of input symbols to which interleaving is applied. The interleaver repeatedly acts with the same algorithm upon successive blocks of L symbols. Often the period of the interleaver is chosen to be equal to the block length of an outer code when block codes are used in serial concatenation. In the concatenation codes, the minimum required *memory* for the employed interleaver equal the length of the processed packet of the encoder *"minimum memory = the codeword length"*.

The idea of reordering the input sequence using the interleaving techniques is spreading the adjacent errors to different places in the transmitted data. So, the probability of burst error within the final received message is decreased. Also, the spreading of bust error leads enhancing the error control schemes correction capability if it is followed by the interleaving techniques [20]. The interleaving techniques are efficient tool to enhance the data transmission over the noisy channel and the mobile communication channel. There are two important parameters of the interleaving techniques, which are the delay of the interleaver and the memory of the interleaver. There is also, the third parameter which determines the efficiency of the interleaver is the "*depth of an interleaver*". Basically, the interleaver technique is utilized for crumbling the adjacent errors. The efficiency of this technique is based on its capability to erasing the burst error with converting it to separated and single errors. Almost, the error spreading due to the interleaving process leads to distributing adjacent errors in the same received codeword to different codewords. So, the capability of error control scheme error correction increases.

The first parameter, the delay is the time required for executing the interleaving process of the input data and the deinterleaving time of the received interleaved data. The delay time of the interleaver and deinterleaver are both equal to the delay caused by the interleaving and deinterleaving process. It should be noted that this time delay is an extra time may increases the overall time of the transmission and receiving the data over the communication networks and the time complexity also.

The second parameter, the amount of storage memory required for the interleaver, this parameter depends on the interleaver size and the input data stream. The efficiency of interleaving techniques related to the optimal delay and memory sense.

6.3.2 Historical Notes

In 1970, Ramsey introduced several types of optimum interleavers, these interleavers optimal sense of minimizing the delay and the memory, it are known his name Ramsey interleavers. In 1971, Forney benefited from the past Ramsey work, he introduced an interleaver called convolutional interleaver. Divsalar and

Pollara presented in 1993 S-random interleaver, it is simple interleaving algorithm. Over the long years of the scientific research, there were several novel approaches to present and design efficient interleavers. The efficient performance of the Turbo code is based on the role of the interleaver utilizing. The researchers focused most of its efforts toward introducing new proposed algorithm of efficient interleaver for application in the Turbo codes [21] or in the interleaved convolutional codes technique [5]. The interleaving techniques also, play an active and important role in the concatenated code schemes where the two Consecutive error control schemes are separated by an interleaver. There are two types of the concatenated codes, serial concatenated codes and parallel concatenated codes, worth mentioning that the turbo code is of the last type [1]. Figure 6.5 shows the block diagram of the serial and parallel concatenated codes.

Interleaving is known as an essential factor influencing performance of turbo codes. In most of designs, a turbo code is implemented as a block code when one block interleaver is used. Another interleaver family, are non-block interleavers, such as convolutional interleavers that have comparable delay with block interleavers, and a simplified implementation. In the case of convolutional interleavers, unlike block interleavers, one input data bit affects the interleaver more than once. Therefore, it is necessary to use continuous decoding methods for such a turbo code. In order to consider a turbo code with the convolutional interleaver as a block code, it is vital to return the interleaver memories to the known state by inserting stuff bits at the end of each block. Block codeword is then created, which makes it possible to implement a conventional iterative decoder that is known as a sub-optimum turbo decoder.

In this chapter, a new version of the interleaving techniques is investigated. This version is based on the idea on traditional data reordering as well as the known encryption tools idea also. The chapter also looked to the structure of the best known block and non-block interleavers are briefly reviewed. Finally, the proposed cooperative algorithm between the WSN codes through the unique secret key based on the presented interleaving chaotic based technique. The proposed algorithm is discussed based on different scenarios of sensing nodes deployment. Also, the chapter contains evaluating of the metrics of the presented scenarios performance.

6.3.3 Interleaving Latency

In the chapter, the interleaving techniques are presented as a tool to enhance the WSN performance with employing new techniques has security capability. So, the latency due to interleaving is a vital factor for the suitability of these techniques for WSN or not. The WSN is real time wireless networks; it transports the real-world and its changes through wireless link to anywhere. Therefore, the delay is not acceptable. Therefore the delay of the captured event is not acceptable, unless leads to reduce the overall delay of the system.

Interleaving generally refers to a process which permutes symbols of an input sequence. It is especially utilized in forward error correction coding to reduce the effect of impulse noise and burst errors in fading and multipath channels. For the same reason it is also applied in magnetic recording systems [22]. Mathematically, the relation between the input sequence $x[k]_{k=-\infty}^{\infty}$ including infinite symbols and the interleaved sequence $y[k] = x[\pi(k)]$ is defined by permuted law $\pi(k)$ as follows [22]:

$$y[k] = x[\pi(k)]$$

An interleaver is considered as a periodic interleaver with the period T when it satisfies the following equation for all symbols of the sequence [22].

$$\pi(x + T) = \pi(x) + T \tag{6.29}$$

Depending on the time difference between an entrance of the symbols to the interleaver and the obtained interleaved sequences, maximum and minimum delays of the interleaver (dmax; dmin) are calculated by [22, 23]:

$$d_{\max} = \max(\pi(x) - x) \tag{6.30}$$

$$d_{\min} = \min(\pi(x) - x) \tag{6.31}$$

Based on the above definitions, the latency parameter is defined as the difference between maximum and minimum delay of the interleaver [7, 22]:

$$d = d_{\max} - d_{\min} \tag{6.32}$$

Also, an interleaver is referred to as a causal interleaver if a symbol cannot exit before it enters. Therefore, a causal interleaver satisfies $d_{\min} \geq 0$ or $d_{\max} \geq d$ relations [7, 22]. For each interleaver, a deinterleaver is defined which reforms the original data at the receiver. Of course, it must be designed to be compatible with the structure of the interleaver. Practically, only interleavers with periodic and causal properties are utilized. Periodic interleavers are mainly subdivided into the block and non-block interleavers. In the block interleavers, permutation of a data block is independently conducted from other blocks. For this purpose, a number of memory units equal to the length of an input sequence is required. The content of every sequence is fully written to the interleaver memories and, according to the permutation law, reading from the memories is accomplished.

Therefore, its delay bound (D) can be assessed by [24, 25]:

$$0 \leq D \leq 2(L - 1) \tag{6.33}$$

where L gives the interleaver length. Those interleavers which delay bound is out of this range, i.e. $D > 2(L - 1)$, are considered as non-block interleavers.

6.4 Interleavers Investigation

In this section, the typical conventional block interleavers applications are introduced. Based on deterministic or random permutation rules, block interleavers can be divided into two main groups. In the deterministic permutation, the positions of each interleaved bit for the whole interleaved data blocks is constant, while in the random permutation these positions are randomly changed. For each permutation rule, relevant interleavers are introduced.

6.4.1 Row-Column Interleaver

The row-column interleaver is a simple block interleaver, which has been utilized in turbo codes. In this interleaver, data are written to its memories row-by-row and then read from them column-by-column [26, 27]. Hence, its permutation is carried out by the following equation:

$$\pi(i) = mn\left[\frac{i}{mn}\right] + n(i\text{mod}n) + \left[\frac{k\text{mod}mn}{m}\right] \tag{6.34}$$

Figure 6.6 shows a general structure of the row-column interleaver. In the case of identical row and column dimensions, the distance between two adjacent bits of the input bitstream in the interleaved data with length L would be $\sqrt{L}$. Although the generated distance is relatively high, in some special cases of turbo code applications, it will not provide sufficiently interleaved data for the second RSC encoder [28, 29]. Figure 6.7 shows the block interleaver behavior with square matrix 8×8, the figure clears the one dimensional burst error and two dimensional burst error.

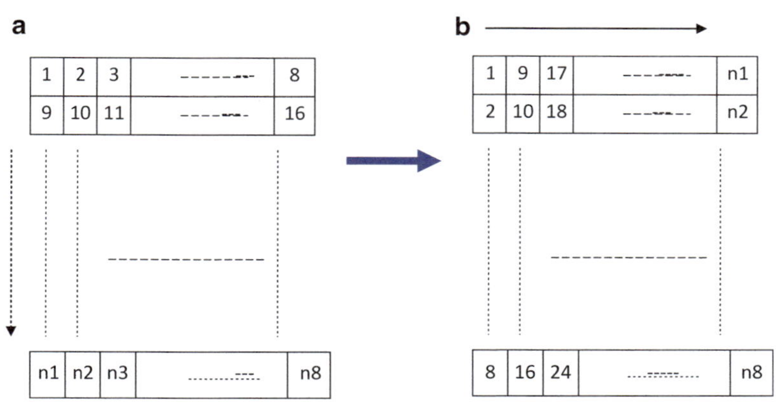

Fig. 6.6 Bit-level interleaver of audio file after transforming to binary data. (**a**) Audio file in binary data form. (**b**) Audio file after bit-level interleaving

a

1	2	3	4	5	6	7	8
9	10	11	12	13	14	15	16
17	18	19	20	21	22	23	24
25	26	26	28	29	30	31	32
33	34	35	36	37	38	39	40
41	42	43	44	45	46	47	48
49	50	51	52	53	54	55	56
57	58	59	60	61	62	63	64

original message

b

1	9	17	25	33	41	49	57
2	10	18	26	34	42	50	58
3	11	19	27	35	43	51	59
4	12	20	28	36	44	52	60
5	13	21	29	37	45	53	61
6	14	22	**30**	**38**	46	54	62
7	15	23	**31**	**39**	47	55	63
8	16	24	32	40	48	56	64

interleaved message

c

1	2	3	4	5	6	7	8
9	10	11	12	13	14	15	16
17	18	19	20	21	22	23	24
25	26	26	28	29	**30**	**31**	32
33	34	35	36	37	**38**	**39**	40
41	42	43	44	45	46	47	48
49	50	51	52	53	54	55	56
57	58	59	60	61	62	63	64

Deinterleaved message

Fig. 6.7 Interleaved data from block interleaver. (**a**) Original message, (**b**) interleaved message and (**c**) deinterleaved message

6.4.2 Reverse Row-Column Interleaver

The reverse row-column interleaver was proposed as an interleaver to remove the drawback of bit 1 existence located at the end part of the interleaver. In this interleaver, data are written to the memories row-by-row and then column-by column reading started from the last column [30, 31]. In a similar way to the row-column interleaver, it cannot remove the problem of bad weight-4 input bit stream. However, it provides good performance for the code with very short block length and for the moderate and long block length at very low signal to noise ratios.

Figure 6.8 shows inverse block interleaver behavior with the same previous square matrix 8×8, the figure clears the one dimensional burst error and two dimensional burst error.

a

1	2	3	4	5	6	7	8
9	10	11	12	13	14	15	16
17	18	19	20	21	22	23	24
25	26	26	28	29	30	31	32
33	34	35	36	37	38	39	40
41	42	43	44	45	46	47	48
49	50	51	52	53	54	55	56
57	58	59	60	61	62	63	64

original message

b

8	16	24	32	40	48	56	64
7	15	23	31	39	47	55	63
6	14	22	30	38	46	54	62
5	13	21	29	37	45	53	61
4	12	20	28	36	44	52	60
3	11	19	**27**	**35**	43	51	59
2	10	18	**26**	**34**	42	50	58
1	9	17	25	33	41	49	57

interleaved message

c

1	2	3	4	5	6	7	8
9	10	11	12	13	14	15	16
17	18	19	20	21	22	23	24
25	**26**	**27**	28	29	30	31	32
33	**34**	**35**	36	37	38	39	40
41	42	43	44	45	46	47	48
49	50	51	52	53	54	55	56
57	58	59	60	61	62	63	64

Deinterleaved message

Fig. 6.8 Interleaved data from reverse block interleaver. (**a**) Original message, (b) interleaved message and (**c**) deinterleaved message

6.4.3 Rotated and Backward Interleaver

Rotated and backward interleavers have been also proposed to remove the effect of bit 1 located at the end part of the row-column interleaver [32, 33]. Similarly to the row-column interleavers, data are written row-by-row. For the rotated interleaver, the written data are rotated 90 and then reading is performed row-by-row. Figure 6.9 shows the interleaved data of the rotated interleaver with the dimension 8×8. In this permutation, the last bit of the input pattern will still be close to the end part of the interleaver. In the backward interleaver, row-by-row reading data starts from the last column of the interleaver in backward direction. Figure 6.10 shows the procedure of backward interleaving. It can be seen that the backward interleaver has

a

1	2	3	4	5	6	7	8
9	10	11	12	13	14	15	16
17	18	19	20	21	22	23	24
25	26	26	28	29	30	31	32
33	34	35	36	37	38	39	40
41	42	43	44	45	46	47	48
49	50	51	52	53	54	55	56
57	58	59	60	61	62	63	64

Original message

b

57	49	41	33	25	17	9	1
58	50	42	34	26	18	10	2
59	51	43	35	27	19	11	3
60	52	44	36	28	20	12	4
61	53	45	37	29	21	13	5
62	54	46	**38**	**30**	22	14	6
63	55	47	**39**	**31**	23	15	7
64	56	48	40	32	24	16	8

Interleaved message

c

1	2	3	4	5	6	7	8
9	10	11	12	13	14	15	16
17	18	19	20	21	22	23	24
25	26	27	28	29	**30**	**31**	32
33	34	35	36	37	**38**	**39**	40
41	42	43	44	45	46	47	48
49	50	51	52	53	54	55	56
57	58	59	60	61	62	63	64

Deinterleaved message

Fig. 6.9 Interleaved data from rotated block interleaver. (**a**) Original message, (**b**) interleaved message and (**c**) deinterleaved message

increased the distance of the last bit of the proposed pattern from the end part of the interleaver. The Figs. 6.9 and 6.10 show the original and interleaved data from modified block interleavers for the Rotated interleaver and the backward interleaver, respectively. Application of these interleavers improves the turbo codes performance especially at medium signal to noise ratios for interleavers with short block lengths.

The backward (I) row-column interleaver is shown in the Fig. 6.10. The reordering of interleaved data seems like the reverse interleaver. The spreading of the proposed received error places is inefficient. The following proposal for backward (II) interleaver starts from the last column from down as given in Fig. 6.11.

a

1	2	3	4	5	6	7	8
9	10	11	12	13	14	15	16
17	18	19	20	21	22	23	24
25	26	26	28	29	30	31	32
33	34	35	36	37	38	39	40
41	42	43	44	45	46	47	48
49	50	51	52	53	54	55	56
57	58	59	60	61	62	63	64

Original message

b

8	16	24	32	40	48	56	64
7	15	23	31	39	47	55	63
6	14	22	30	38	46	54	62
5	13	21	29	37	45	53	61
4	12	20	28	36	44	52	60
3	11	19	**27**	**35**	43	51	59
2	10	18	**26**	**34**	42	50	58
1	9	17	25	33	41	49	57

Interleaved message

c

1	2	3	4	5	6	7	8
9	10	11	12	13	14	15	16
17	18	19	20	21	22	23	24
25	**26**	**27**	28	29	30	31	32
33	**34**	**35**	36	37	38	39	40
41	42	43	44	45	46	47	48
49	50	51	52	53	54	55	56
57	58	59	60	61	62	63	64

Deinterleaved message

Fig. 6.10 Interleaved data from backward (I) block interleaver. (**a**) Original message, (**b**) interleaved message and (**c**) deinterleaved message

6.4.4 Convolutional Interleaver Structure

The structure of the convolutional interleaver is similar to the Ramsey Type *I* interleaver [24, 25]. This interleaver is constructed by T parallel lines, which defines its period. Conventionally, each interleaver line has a different number of memories from the other lines. The difference in the numbers of memories between two adjacent interleaver lines is generally considered as a constant referred to as a space parameter of the interleaver [34, 35]. Figure 6.12 illustrates the general structure of convolutional interleavers and deinterleavers with period T and space value $M=1$. The number of memories in each line of the deinterleaver is determined based on the number of memories applied at the corresponding interleaver

a

1	2	3	4	5	6	7	8
9	10	11	12	13	14	15	16
17	18	19	20	21	22	23	24
25	26	26	28	29	30	31	32
33	34	35	36	37	38	39	40
41	42	43	44	45	46	47	48
49	50	51	52	53	54	55	56
57	58	59	60	61	62	63	64

Original message

b

64	56	48	40	32	24	16	8
63	55	47	39	31	23	151	7
62	54	46	38	30	22	14	6
61	53	45	37	29	21	13	5
60	52	44	36	28	20	12	4
59	51	43	**35**	**27**	19	11	3
58	50	42	**34**	**26**	18	10	2
57	49	41	33	25	17	9	1

Interleaved message

c

1	2	3	4	5	6	7	8
9	10	11	12	13	14	15	16
17	18	19	20	21	22	23	24
25	**26**	**27**	28	29	30	31	32
33	**34**	**35**	36	37	38	39	40
41	42	43	44	45	46	47	48
49	50	51	52	53	54	55	56
57	58	59	60	61	62	63	64

Deinterleaved message

Fig. 6.11 Interleaved data from backward (II) block interleaver. (**a**) Original message, (**b**) interleaved message and (**c**) deinterleaved message

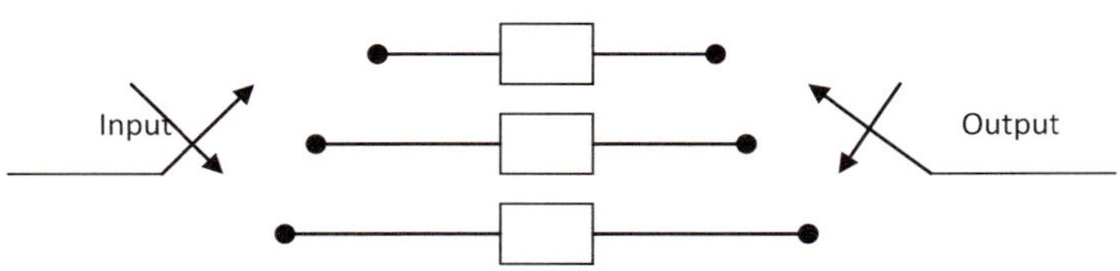

Fig. 6.12 Convolutional interleaver structures

line, such that distributed information in every corresponding interleaver and deinterleaver line will together pass through to the same number of memories. Also, the selectors applied for these structures have to synchronously operate with each other to properly recover the original data at the end of decoder. Based on the arithmetic sequence, the overall number of memories for the interleaver (*T*, *M*) is given by [36, 37]:

$$\begin{aligned} S = \sum_{i=1}^{T} s_i &= M + 2M + \cdots\cdots + (T-1)M \\ &= \frac{T(T-1)M}{2} \end{aligned} \tag{6.35}$$

A convolutional interleaver can be forced to operate as a block interleaver by inserting a number of zero stuff bits to its memories providing an interleaved data block which is isolated from the other blocks [38]. Initially, the interleaver memories are set to zero values. Based on the number of applied memories in each line, bits distributed to the relevant line appear at different times. For example, the interleaved data block of an input bitstream with the length $L = 12$ from the interleaver ($T = 3$; $M = 1$) is given by:

$$\{x_0, o, o, x_3, x_1, o, x_6, x_4, x_2, x_9, x_7, x_5, o, x_{10}, x_8, o, o, x_{11}\}.$$

Depending on the input sequence length and the interleaver parameters, distributed data are terminated at one of the interleaver lines, which is determined by the Re $m(L, T)$ value, where Rem(L; T) gives the reminder of *L*/*T* operation. For an input bits-stream $\{x_0, x_1, x_2, \cdots, x_L\}$ with the length L, different interleaved data for the interleaver (T = 3; M = 1) would be:

$$\begin{aligned} &\text{Rem}(L, T) = 0: \\ &\{x_0, o, o, x_3, x_1, o, x_6, \cdots, x_{L-7}, o, x_{L-2}, x_{L-4}, o, o, x_{L-1}\} \\ &\text{Rem}(L, T) = 1: \\ &\{x_0, o, o, x_3, x_1, o, x_6, x_4, \cdots, x_{L-3}, x_{L-5}, o, o, x_{L-2}\} \\ &\text{Rem}(L, T) = 2: \\ &\{x_0, o, o, x_3, x_1, o, x_6, \cdots, x_{L-4}, x_{L-6}, o, x_{L-1}, x_{L-3}\} \end{aligned}$$

When a convolutional interleaver with the above structure is applied in the turbo code structure, inserted stuff bits to the interleaver memories reduce channel bandwidth usage. Therefore, an optimization can be performed on the interleaver to control the number of those bits equal to the number of applied memories. For this purpose, one block is added after the interleaver controlling the data at the interleaver output in order to delete extra zero stuff bits that appear at the end part of the interleaver [38]. In this case, the memory contents at the end of each block have zero value. This remains until the beginning of the next block. Following the previous example, the optimized interleaver outputs for different Rem(L, T) values are given by:

$$\text{Rem}(L,T)=0:$$
$$\{x_0,o,o,x_3,x_1,o,x_6,\cdots,x_{L-2},x_{L-4},x_{L-1}\}$$
$$\text{Rem}(L,T)=1:$$
$$\{x_0,o,o,x_3,x_1,o,x_6,x_4,\cdots,x_{L-3},x_{L-5},x_{L-2}\}$$
$$\text{Rem}(L,T)=2:$$
$$\{x_0,o,o,x_3,x_1,o,x_6,\cdots,x_{L-4},x_{L-6},x_{L-1},x_{L-3}\}$$

Cases of the convolutional interleaver with the sequence of x:

```
############:
>>x11=[1 2 3 4 5 6 7 8 9 10 11 12 13 14 15 16 17 18 19 20 21 22 23 24 25 26 27 28
29 30 31 32 33 34 35 36 37 38 39 40 41 42 43 44 45 46 47 48 49 50 51 52 53 54
55 56 57 58 59 60 61 62 63 64 0 0 0 0 0 0];
>>delay=[0 1 2]; % Set delays of three shift registers.
>>[y11,state_y]=muxintrlv(x11,delay); % Interleaved message.

y11=[1 0 0 4 2 0 7 5 3 10 8 6 13 11 9 16 14 12 19 17 15 22 20 18 25 23 21 28 26 24
31 29 27 34 32 30 37 35 33 40 38 36 43 41 39 46 44 42 49 47 45 52 50 48 55 53
51 58 56 54 61 59 57 64 62 60 0 0 63 0]

>>z11=muxdeintrlv(y11,delay) % Deinterleaved message

z11=[0 0 0 0 0 0 x]
```

The zeros within interleaved message are ignored in the evaluating example. This example of the interleaved and deinterleaved message using the convolutional interleaver with (T = 4 and m = 1) is evaluated using the simple matrix of x input (1–64), x represents “sample of data or bits” image pixels or audio samples. The original and interleaved matrix is presented in this form to propose the two cases of burst errors, one dimensional burst error and two dimensional burst errors, also, what interleaver impact in the two presented cases will be cleared. The previous cases of the same message and the same burst error pattern are presented with different types of row-column block interleavers. The input of the Convolutional interleaver is the same 8×8 square matrix. The performance of this interleaver is cleared and given in Fig. 6.13.

With the message sequence x1 = [0 0 × 0 0 0 0 0 0], the interleaved message and the deinterleaved message with the spreading of the burst errors pattern is shown in Fig. 6.14. The result of interleaved matrix from the convolutional interleaver as shown in Fig. 6.14 shows the effectiveness of this interleaver for breaking the adjacent errors and the two dimensional burst errors but with limited separation. This ability of the convolutional interleaver leads to decreasing the effect of burst error appearance but it does not reach to raise the ability of error control scheme in

a

1	2	3	4	5	6	7	8
9	10	11	12	13	14	15	16
17	18	19	20	21	22	23	24
25	26	26	28	29	30	31	32
33	34	35	36	37	38	39	40
41	42	43	44	45	46	47	48
49	50	51	52	53	54	55	56
57	58	59	60	61	62	63	64

Original message

b

1	4	2	7	5	3	10	8
6	13	11	9	16	14	12	19
17	15	22	20	18	25	23	21
28	26	24	31	29	27	34	32
30	37	35	33	40	38	36	43
41	39	46	44	42	49	47	45
52	50	48	55	53	51	58	56
54	61	59	57	64	62	60	63

Interleaved message

c

1	2	3	4	5	6	7	8
9	10	11	12	13	14	15	16
17	18	19	20	21	22	23	24
25	26	27	28	29	30	31	32
33	34	35	36	37	38	39	40
41	42	43	44	45	46	47	48
49	50	51	52	53	54	55	56
57	58	59	60	61	62	63	64

Deinterleaved message

Fig. 6.13 Interleaved data from convolutional interleaver-I. (**a**) Original message, (**b**) interleaved message and (**c**) deinterleaved message

case of using it within the interleaved FEC technique, where the separation between the distributed error let it located in the same codeword.

6.5 Chaotic Interleaver Structure

In this section and the following subsection, the interleaving chaotic based is discussed. The presented technique can be considered the newest version to modernize and novelize the known interleaver. The expected performance of the interleaver chaotic based is powerful than the traditional ones. There are little researchers merged the interleaving technique chaotic based on the different

a

1	2	3	4	5	6	7	8
9	10	11	12	13	14	15	16
17	18	19	20	21	22	23	24
25	26	26	28	29	30	31	32
33	34	35	36	37	38	39	40
41	42	43	44	45	46	47	48
49	50	51	52	53	54	55	56
57	58	59	60	61	62	63	64

Original message

b

2	5	3	1	8	6	4	11
9	7	14	12	10	17	15	13
20	18	16	23	21	19	26	24
22	29	27	25	32	30	28	35
33	31	38	36	34	41	39	37
44	42	40	**47**	**45**	43	50	48
46	53	51	**49**	**56**	54	52	59
57	55	62	60	58	63	61	64

Interleaved message

c

1	2	3	4	5	6	7	8
9	10	11	12	13	14	15	16
17	18	19	20	21	22	23	24
25	26	27	28	29	30	31	32
33	34	35	36	37	38	39	40
41	42	43	44	**45**	46	**47**	48
49	50	51	52	53	54	55	**56**
57	58	59	60	61	62	63	64

Deinterleaved message

Fig. 6.14 Interleaved data from convolutional interleaver-I. (**a**) Original message, (**b**) interleaved message and (**c**) deinterleaved message

systems. The starting of this version of interleaving begins in 2004 by Zhang et al. It utilized as an interleaver within the turbo code. It was based on simple one dimensional chaotic map. In 2010, it merged in BICM technique using the same chaotic map by Xuelan [39]. Also, Emad et al. in 2012 employed it in the OFDM system. The original work of the chaotic maps is data encryption tool. Nevertheless, many of researchers studied its applying on different systems such as the interleaved error control techniques, Turbo code, OFDM, and etc.

6.5.1 Idea of Chaotic Interleaver

The tools of data encryption depend on randomizing the transmitted data using secret key. The idea of utilizing the power of data randomization of these techniques is work as an interleaver. There are two types of chaotic maps, one dimensional chaotic map and two dimensional chaotic maps, in general the chaotic map techniques is used as encryption techniques in image transmission. It is used to randomize the image pixels.

Zhang et al. [38] present a Block-Random Chaotic (BRC) interleaver, which has lower complexity than the semi-random interleaver with the similar performance. Application of this interleaver has been verified for the 3rd Generation (3G) of mobile communications where the chaotic sequence is considered as input bit stream of the turbo code. The chaotic interleaver idea is presented using the logistic map in [1, 5, 38]. This map is defined as a typical example of a 1-D chaotic map. In [5], it is applied with the turbo code. In [1], a logistic interleaver is employed in a Bit-Interleaved Coded Modulation (BICM) technique. In this work, the proposed chaotic interleaver is based on a 2-D Baker map, which is a powerful data randomization tool. This map is widely used as an encryption tool [2, 3], and it will be developed in this chapter as an interleaving tool.

6.5.2 Cryptography with Chaotic System

The chaotic system is data randomizing tool. It is almost known as an encryption tool. In general, it can be said that the robustness of ciphertext using the chaotic system higher than the traditional interleavers. Against noise A well designed encryption system must guarantee the diffusion property, i.e., a small change in either the key or the plaintext must lead to a totally different ciphertext. Therefore if the ciphertext is slightly modified, then the decryption process should return a plaintext totally different from the original one. The authors of [30] claim that their cryptosystem implements a good diffusion procedure. At the same time, they also claim that the cryptosystem is robust against noise, which contradicts the previous assertion about 4 diffusion. In order to clarify this point, we performed some simulations where the ciphertext was modified. An efficient cryptosystem must allow the perfect recovery of the plaintext from the ciphertext when the key is known.

6.5.3 Definition and Goals of Cryptography

Cryptography is a study of techniques (called cryptosystems) that are used to accomplish the following four goals: Confidentiality, Data Integrity, Authentication and Non-repudiation.

A study of techniques used to break existing cryptosystems is called ***Cryptanalysis***. Since cryptography and cryptanalysis are greatly dependent of each other, people refer to ***Cryptology*** as a joint study of cryptography and cryptanalysis. Now Let us try to understand all four goals of cryptography.

Confidentiality refers to the protection of information from unauthorized access. An undesired communicating party, called adversary must not be able to access the communication material. This goal of cryptography is a basic one, one that has been always addressed and enforced throughout the history of cryptographic practice.

Data integrity ensures that information has not been manipulated in an unauthorized way. If the information is altered, all communicating parties can detect

Authentication methods are studied in two groups: Entity authentication and message authentication. Entity authentication is also often referred to as entity identification or only identification. Message authentication is closely related to entity authentication and recognition. Entity authentication is the process whereby one party is assured (through acquisition of corroborative evidence) of the identity of a second party involved in a protocol, and that the second has actually participated (i.e., is active at, or immediately prior to, the time the evidence is acquired). Message authentication is a term used analogously with data origin authentication. It provides data origin authentication with respect to the original message source (and data integrity, but no uniqueness and timeliness guarantees). The difference between message authentication and entity authentication is the lack of a timeliness guarantee in the former with respect to when a message was created. This follows from the fact that providing a message authentication gives no guarantee that the message authentication was created immediately prior to sending the message. Thus a major difference between entity authentication and message authentication is that entity authentication involves two parties communicating actively. To provide this requirement an entity authentication scheme has to involve some kind of clock or timeliness.

Non-repudiation means that the receiver can prove to everyone that sender did indeed send the message; i.e., the sender cannot claim that he or she didn't encrypt and/or sign certain digital information.

Fortunately, modern cryptography has developed techniques to handle all four goals of cryptography.

Principles of Encryption

The basic idea of ***Encryption*** is to modify the message in such a way that only a legal recipient can reconstruct its content. A discrete-valued cryptosystem can be characterized by:

A set of possible plaintexts, P.
A set of possible ciphertexts, C.
A set of possible cipher keys, K.
A set of possible encryption and decryption transformations, E and D.

An encryption system is also called ***a cipher***, or a cryptosystem. The message for encryption is called ***plaintext***, and the encrypted message is called ***ciphertext***. Denote the plaintext and the ciphertext by P and C, respectively. The encryption procedure of a cipher can be described as:

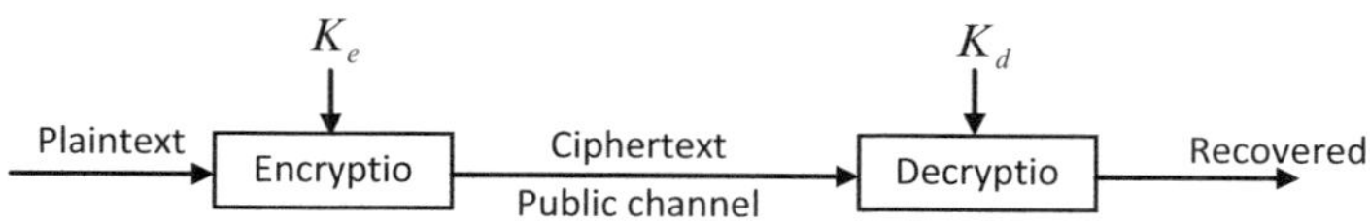

Fig. 6.15 Encryption/decryption of a cipher

$$C = E_{K_e}(P) \tag{6.36}$$

where K_e is the encryption key and $\boldsymbol{E}$ is the encryption function. Similarly, the decryption procedure is defined as:

$$P = D_{K_d}(C) \tag{6.37}$$

where K_d is the decryption key and $\boldsymbol{D}$ is the decryption function. The security of a cipher should only rely on the decryption key K_d, since adversaries can recover the plaintext from the observed ciphertext once, it gets K_d. Figure 6.15 shows a block diagram for encryption/decryption of a cipher.

6.5.4 Encryption Classification Algorithms

Encryption algorithms, also called ciphers, can also be classified by different ways, such as with respect to the structures of the algorithms, or with respect to keys, or with respect to the approach in constructing the scheme, or with respect to the percentage of the data encrypted.

a. Classification according to encryption structure

Encryption algorithms can be classified according to encryption structure into block ciphers and stream ciphers.

A block cipher is a type of symmetric-key encryption algorithm that transforms a fixed-length block of plaintext data into a block of ciphertext data of the same length. The fixed length is called the block size, and for many block ciphers, the block size is 64 or 128 bits. The larger the block size, the more secure the cipher, but the more complex encipher and decipher algorithms and devices. Typical block ciphers include DES, Triple TDES, RC5, RC6, Blowfish, IDEA and Rijndael. Some of them have become standard cipher lately. Modern block ciphers have the following features:

1. Variable key length.
2. Mixed arithmetic operations, this can provide non-linearity.
3. Data-dependent rotations and key-dependent rotations.
4. Lengthy key schedule algorithm.
5. Variable plaintext/ciphertext blocks length and variable number of rounds.

Block ciphers can be characterized by:

1. Block size: larger block sizes mean greater security.
2. Key size: larger key sizes mean greater security.
3. Number of rounds: multiple rounds offer increasing security.
4. Encryption modes: define how messages larger than the block size are encrypted.

Unlike block ciphers that operate on large blocks of data, ***Stream ciphers*** typically operate on smaller units of plaintext, usually bits. So, stream ciphers can be designed to be exceptionally fast, much faster than a typical block cipher. Generally, a stream cipher generates a sequence of bits as a key (called keystream) using Pseudo Random Number Generator (PRNG) that expand a short secret key (e.g., 128-bit) into a long string (keystream) (e.g., 10^6 bit), and the encryption is accomplished by combining the keystream with the plaintext. Usually, the bitwise XOR operation is chosen to perform ciphering, basically for its simplicity. Stream ciphers have the following properties:

1. It does not have perfect security.
2. Security depends on properties of PRNG.
3. PRNG must be unpredictable; given consecutive sequence of bits output, next bit must be hard to predict.
4. Typical stream ciphers are very fast

As of today, there is no stream cipher that has emerged as a standard. The most widely used stream cipher is RC4. It may be employed in many places such as in SSL (RC4), DVD (LFSR), WEP (RC4), etc.

b. Classification according to keys

According to keys, there are two kinds of ciphers following the relationship of K_e and K_d. When $K_e = K_d$, the cipher is called a private-key cipher or a symmetric cipher. For private-key ciphers, the encryption/decryption key must be transmitted from the sender to the receiver via a separate secret channel. When $K_e \neq K_d$, the cipher is called a public-key cipher or asymmetric cipher. For public-key ciphers, the encryption key K_e is published and the decryption key K_d is kept private, for which no additional secret channel is needed for key transfer.

In conventional encryption as shown in Fig. 6.16, the sender encrypt the data (plaintext) using the encryption key and the receiver decrypt the encrypted data (ciphertext) into original data (plaintext) using the decryption key. In symmetric encryption both encryption and decryption keys are identical.

Figure 6.17 shows the public key encryption (asymmetric encryption) in which the encryption and decryption keys are different, instead of one key there are two different keys, one is public (K_e) and the other is private (K_d). Public key cryptography solves the problem of conventional cryptosystem by distributing the key. Table 6.2 shows a comparison between symmetric encryption and asymmetric encryption.

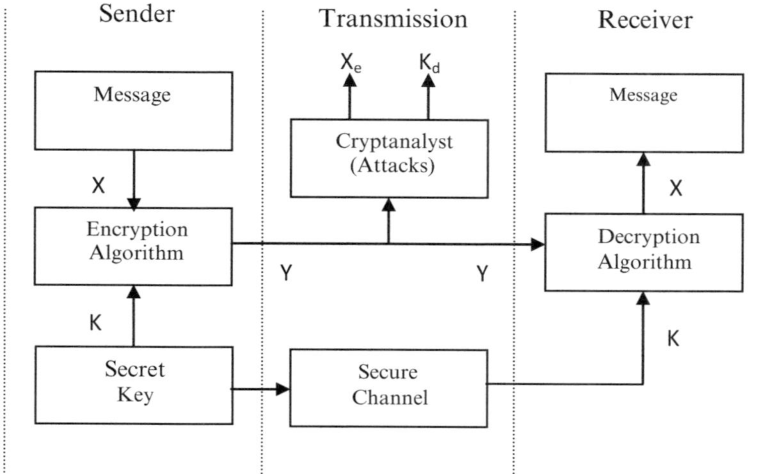

Fig. 6.16 Model of symmetric encryption

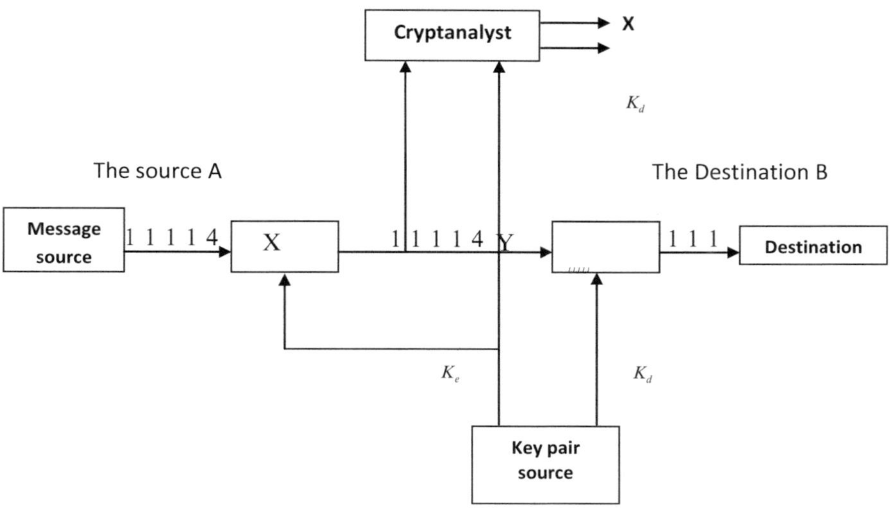

Fig. 6.17 Public key encryption

In general, there are two types of cryptosystems:

1. Symmetric (private) key cryptosystems.
2. Asymmetric (public) key cryptosystems.

Most people have chosen to call the first group simply symmetric key cryptosystems, and the popular name for the second group is just public key cryptosystems.

Table 6.2 Comparison between symmetric encryption and asymmetric encryption

Conventional encryption symmetric encryption	Public-key encryption asymmetric encryption
Needed to work:	*Needed to work*:
1. The same algorithm with the same key can be used for encryption and decryption	1. One algorithm is used for encryption and decryption with a pair of keys, one for encryption and one for decryption
2. The sender and receiver must share the algorithm and the key	2. The sender and receiver must each have one of the matched pair of keys
Needed for security:	*Needed for security*:
1. The key must be kept secret	1. One of the two keys must be kept secret
2. It must be impossible or at least impractical to decipher a message if no other information is available	2. It must be impossible or at least impractical to decipher a message if no other information is available
3. Knowledge of the algorithm plus samples of cipher text must be insufficient to determine the key	3. Knowledge of the algorithm plus one of the keys plus samples of cipher text must be insufficient to determine the key

6.6 Cryptography Using Chaos

Chaos theory has been established since 1970s by many different research areas, such as physics, mathematics, engineering, and biology, etc.

The most well-known characteristics of chaos are so called "butterfly-effect" (sensitivity to initial conditions), and the pseudo-randomness generated by deterministic equations. Many researchers have noticed that there exists an interesting relationship between chaos and cryptography. Many fundamental properties of chaotic systems have their corresponding counterparts in traditional cryptosystems. Chaotic systems have several significant features favorable to secure communications, such as ergodicity, sensitivity to initial condition, control parameters and random like behavior, which can be connected with some conventional cryptographic properties of good ciphers, such as confusion/diffusion.

With all these advantages, scientists expected to introduce new and powerful tools of chaotic cryptography. So, Chaos may become a new rich source of new ciphers. We use the chaotic characteristic for encrypting image to generate good cipherimage. Below, we will make a description for chaos, including the properties of chaos, one-dimensional chaotic logistic map, and the relationship between chaos and cryptography.

6.6.1 The Properties of Chaos

The main basic properties of chaotic systems are deterministic, the sensitivity to initial conditions and parameters, ergodicity and mixing (topological transitivity).

Deterministic, means that they have some determining mathematical equations ruling their behavior.

The sensitivity to initial conditions (butterfly effect) means that when a chaotic map is iteratively applied to two initially close points, it iterates quickly diverge, and become uncorrelated in the long term. Sensitivity to parameters causes the properties of the map to change quickly when slightly perturbing the parameters on which the map depends. Hence, a chaotic system can be used as a pseudo-random number generator.

Mixing, or topological transitivity, is the tendency of the system to quickly "scramble up" small portions of the state space into an intricate network of filaments. Local, correlated information becomes scattered all over the state space.

The ergodicity property a chaotic map means that if partitioning the state space into a finite number of regions, no matter how many, any orbit of the map will pass through all these regions. The ergodicity of a chaotic map is ensured by the topological transitivity property

Chaotic map's properties are in close relation with the cryptosystem's security. At first, its parameter is used as confusion key. Thus, parameter sensitivity is in close relation with key sensitivity. The higher the parameter sensitivity, the higher the key sensitivity, and the stronger the cryptosystem. Thus, the chaotic map with high parameter sensitivity is preferred in this cryptosystem. Secondly, the initial-value sensitivity and state ergodicity of the chaotic map determine the confusion strength. In chaotic confusion process, initial value refers to the initial position of a pixel. Thus, the higher the initial-value sensitivity, the smaller the correlation between adjacent pixels, and the more random the confused image. Similarly, state ergodicity means that a pixel in a certain position can be permuted to any position with the same probability. Thus, the higher the state ergodicity, the more random the confusion process, and the more difficult the statistical attack. Therefore, the chaotic map with high initial-value sensitivity and state ergodicity is preferred in this cryptosystem.

For diffusion function, a change of a pixel can spread to other pixels, which keeps the cryptosystem of high plaintext-sensitivity. The more the pixels are changed in one round of the diffusion process, the higher the diffusion speed of the diffusion function will be. Considering that different diffusion function has different diffusion speed, the one with high diffusion speed is preferred in this cryptosystem.

Besides chaotic map and diffusion function, iteration time is in close relation with cryptosystem's security. The more the iteration time, the larger the cryptosystem's key space is if different keys are used in different iteration.

Also, the more the iteration time is, the higher the cryptosystem's encryption strength is. So bigger key space is recommended in order to obtain high security. The general properties of chaotic maps guarantee: (a) cipher security, (b) very large number of ciphering keys, (c) easy parallelization, (d) simple and fast implementation.

6.6.2 One Dimensional Chaotic Map

To introduce one-dimensional chaotic maps, the logistic map is given by as:

$$X_{n+1} = rX_n \ (1 - X_n) \tag{6.38}$$

Where X_n takes values in the interval [0,1]. It is one of the simplest models that present chaotic behavior. The parameter r is a positive constant and takes values up to 4. Its value determines and explores the behavior of the logistic map.

6.6.3 Two Dimensional Chaotic Map

Chaotic encryption of the watermark image is performed using the chaotic Baker map. The Baker map is a chaotic map that generates a permuted version of a square matrix. In its discretized form, the Baker map is an efficient tool to randomize a square matrix of data. The discretized map can be represented for an R × R matrix as follows:

$$B(r_1, r_2) = \left[\frac{R}{n_i}(r_1 - R_i) + r_2 \mathrm{mod}\left(\frac{R}{n_i}\right), \frac{n_i}{R}\left(r_2 - r_2 \mathrm{mod}\left(\frac{R}{n_i}\right)\right) + R_i\right] \tag{6.39}$$

where $B(r_1, r_2)$ are the new indices of the data item at (r_1, r_2), $R_i \leq r_I \leq R_i + n_i$, $0 < r_2 < R$, and $R_i = n_I + n_2 + \ldots + n_i$.

In steps, the chaotic encryption is performed as follows:

- An $R \times R$ square matrix is divided into R rectangles of width n_i and number of elements R.
- The elements in each rectangle are rearranged to a row in the permuted rectangle. Rectangles are taken from right to left beginning with upper rectangles then lower ones.
- Inside each rectangle, the scan begins from the bottom left corner towards upper elements.

Figure 6.18 shows an example for the chaotic encryption of an 8 × 8 square matrix (i.e. $R = 8$). The secret key is $S_{key} = [n_1, n_2, n_3] = [2, 4, 2]$.

In [6] the selection of the chaotic encryption algorithm for watermark encryption is attributed to its permutation nature, which resists noise of the channel and other attacks effect. After watermark embedding, the audio signal is transformed again into a 1-D format. The transformation between the 1-D and 2-D formats is performed.

Original square matrix. Chaotic encrypted matrix.

Fig. 6.18 Chaotic encryption of an 8 × 8 matrix

6.6.4 *Chaos and Cryptography Relationship*

There have been many discussions in the literature about the relationships between chaotic systems and cryptosystems. Many properties of chaotic systems have their corresponding counterparts in traditional cryptosystems. Table 6.3 contains a partial list of these properties.

A deep relation between chaos and cryptography has not been established yet. The main difference between chaos theory and cryptography is that cryptosystems work on a finite field, while chaos is meaningful only on a continuum. Nevertheless, these two scientific notions do have very tight relationships. Many fundamental concepts in chaos theory, such as mixing and sensitivity to initial conditions and parameters, actually coincide with those in cryptography. The similarities and differences between the two subjects can be listed out, as shown in Fig. 6.19.

Form Fig. 6.19, it is clear that chaotic maps and cryptographic algorithms have some similar properties: both are sensitive to a tiny change in initial conditions and parameters; both have random-like behaviors; cryptographic algorithms achieve confusion and diffusion data by rounds of encryption, while chaotic maps spread a small region of data over the entire phase space via iterations. The only difference in this regard is that encryption operations are defined on finite sets of integers while chaos, on real numbers. Moreover, for the time being, the notions of cryptographic security and performance of cryptographic algorithms have no counterpart in chaos theory.

Table 6.3 Comparison between chaos and cryptography properties

Chaotic property	Cryptographic property	Description
Ergodicity	Confusion	The output has the same distribution for any input
Sensitivity to initial conditions/control parameter	Diffusion with a small change in the plaintext/secret key	A small deviation in the input can cause a large change at the output
Mixing property	Diffusion with a small change in one plain-block of the whole plaintext	A small deviation in the local area can cause a large change in the whole space
Deterministic dynamics	Deterministic pseudo-randomness	A deterministic process can cause a random-like (pseudo-random) behavior
Structure complexity	Algorithm (attack) complexity	A simple process has a very high complexity

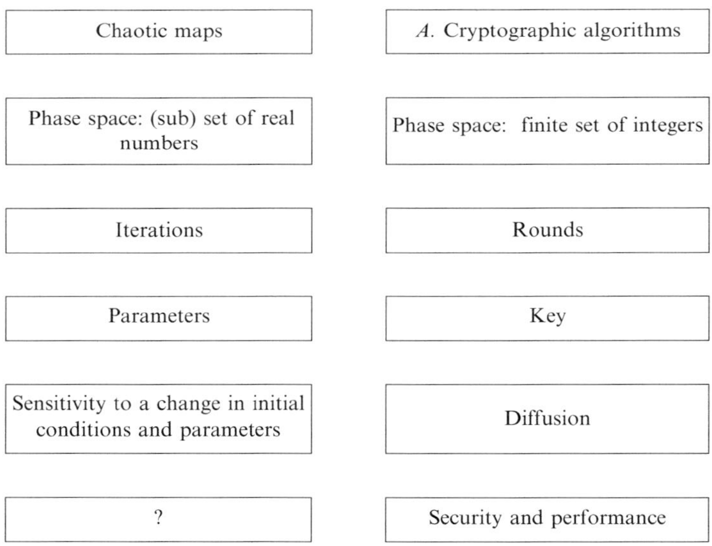

Fig. 6.19 Similarities and differences between chaotic systems and cryptographic algorithms

6.6.5 *Chaotic Interleaver*

In the following, the two dimensional chaotic Baker map based is utilized as an efficient interleaving technique. The previous sections show the performance of traditional interleaver with the one dimensional burst error and two dimensional burst errors. The goal of utilizing the encryption tools as an interleaver can be cleared from the following figure. The results of this figure are gotten by the Eq. 6.39. The performance of this interleaver is shown in Fig. 6.20.

a

1	2	3	4	5	6	7	8
9	10	11	12	13	14	15	16
17	18	19	20	21	22	23	24
25	26	26	28	29	30	31	32
33	34	35	36	37	38	39	40
41	42	43	44	45	46	47	48
49	50	51	52	53	54	55	56
57	58	59	60	61	62	63	64

Original message

b

31	23	15	7	32	24	16	8
63	55	47	39	64	56	48	40
11	3	12	4	13	5	14	6
27	19	28	20	29	21	30	20
43	35	44	36	45	37	46	38
59	51	60	**52**	**61**	53	62	54
25	17	9	**1**	**26**	18	10	2
57	49	41	33	58	50	42	34

Interleaved message

c

1	2	3	4	5	6	7	8
9	10	11	12	13	14	15	16
17	18	19	20	21	22	23	24
25	**26**	27	28	29	30	31	32
33	34	35	36	37	38	39	40
41	42	43	44	45	46	47	48
49	50	51	**52**	53	54	55	56
57	58	59	60	**61**	62	63	64

Deinterleaved message

Fig. 6.20 Interleaved data from chaotic interleaver. (**a**) Original message, (**b**) interleaved message and (**c**) deinterleaved message

6.7 Objective Quality Metrics

In this section, the different metrics are devoted to evaluate the presented scenarios for WSN performance with image transmission. The transmitted image over the mobile communication channel is very sensitive to the noise more than the fixed channel. The metrics are various to measure the received image samples quality and the reliability of the link also through the Bit Error Rate (BIR) and the Number of Lost packets (NLP). First metric is the Peak Signal to Noise ratio (PSNR), it is utilized to measure the quality of received image samples. The PSNR of the received images is used as an evaluation metric in this scenario. The Mean Square

Error (MSE) is the cumulative squared error between the received image and the original image. The mathematical formula for MSE is given in Eq. 6.40 [23].

$$MSE = \frac{1}{MN}\sum_{y=1}^{M}\sum_{x=1}^{N}\left[I(x,y) - I'(x,y)\right]^2 \tag{6.40}$$

$$PSNR = 20x\ \log_{10}(255/sqrt(MSE)) \tag{6.41}$$

where $I(x, y)$ is the original image, $I'(x, y)$ is the received image, and M, N are the image dimensions. The higher PSNR is good that means the SNR is higher. Also, the case of lower MSE means lesser error, from Eq. 6.40 there is the inverse relation between the MSE and PSNR. The best technique is having a lower MSE and higher PSNR [25].

The second metric is used in this scenario is measuring the degree of the correlation between the original and received images as given in Eq. 6.42 for different cases of our simulation [40].

$$C_r = corr(I(x,y), I'(x,y)) \tag{6.42}$$

C_r is correlation coefficient, $C_r = 1$ in the case of perfect correlation. As it approaches zero there is less of a relation, which means closer to uncorrelated.

The number of lost packet percentage is the metric of wireless channel reliability. The reliability of the link can be defined by the difference between the successes and corrected received packets and the total transmitted packets from the SN to the sink. The reliability R is given by the percentage of the sent packets that arrive correctly at the sink node and it may be evaluated as:

$$R = \frac{P_C}{P_S} \tag{6.43}$$

$$P_C = P_S - P_E \tag{6.44}$$

$$R = \frac{P_S - P_E}{P_s} \tag{6.45}$$

P_C is the number of corrected packets, P_S is the total of transmitted packets. P_E is the number of dropped packets.

The NLP percentage is the number of failed packet to the total number of transmitted packets from the node to the sink [31].

$$NLP\% = \frac{P_S - P_C}{P_S} \times 100 \tag{6.46}$$

So, with lower NLP means higher reliability of link. It means also, lower retransmission request times and power efficient link [41, 42].

BER = No. of bit error/length of packet*N iteration

PER = No. of dropped packets/Total Send packet, the dropped packet ratio to the total send packets.

NLP = Total send packet – success received packets.

NLP% = NLP/Total*100

Also, from the comparison purpose at different SNRs, the number of lost frames is studied with the channel SNR. In our work, MATLAB was used for carrying out the simulation experiments of different cases. The simulation results have been gotten by transmission of the image over different SNR values [9].

6.8 Jakes Method

This model is an accepted model utilized for simulate the correlated fading channel. The Jakes fading model, also known as the Sum of Sinusoids model, is a deterministic method for simulating time-correlated Rayleigh fading waveforms and is still widely used today. The model assumes that N equal-strength rays arrive at a moving receiver with uniformly distributed arrival angles α_n, such that ray n experiences a Doppler shift ω_n, $\omega_n = \omega_m \cos(\alpha_n)$, where $\omega_m = 2\pi fv/c$ is the maximum Doppler frequency shift, v is the vehicle speed, f is the carrier frequency, and c is the speed of light [27]. Using $\alpha_n = 2\pi n/\mathrm{N}$, there is quadrantal symmetry in the magnitude of the Doppler shift, except for angles 0 and π. As a result, the fading waveform can be modeled with $No + 1$ complex oscillators, where $No = (N/2 - 1)/2$. This leads to the equation,

$$T_k(t) = \sqrt{\frac{1}{2N_o + 1}} \left\{ 2 \sum_{n=1}^{N_o} (\cos\beta_n + j\sin\beta_n)\cos(\omega_m \cos\alpha_n t + \theta_{nk}) + \sqrt{2} + \cos(\omega_m t + \theta_{0k}) \right\} \tag{6.47}$$

where, k is the waveform index, $k = 1,2,\ldots.N0$ and λ is the wavelength of the transmitted carrier frequency. Here, $\beta_n = \pi n/(N_0 + 1)$. To generate the multiple waveform, Jakes suggests using

$$\theta_{nk} = \frac{\pi n}{N_0 + 1} + \frac{2\pi(k-1)}{N_0 + 1} \tag{6.48}$$

A software simulation was carried out using the Jakes Model described above and the inputs included the Vehicle Speed v (in kmph), the Central Carrier Frequency f_c in MHz, the Symbol Frequency fs in kbps, the Number of sub-channels U and the Number of Channel Coefficients to Generate M. The output was shown as

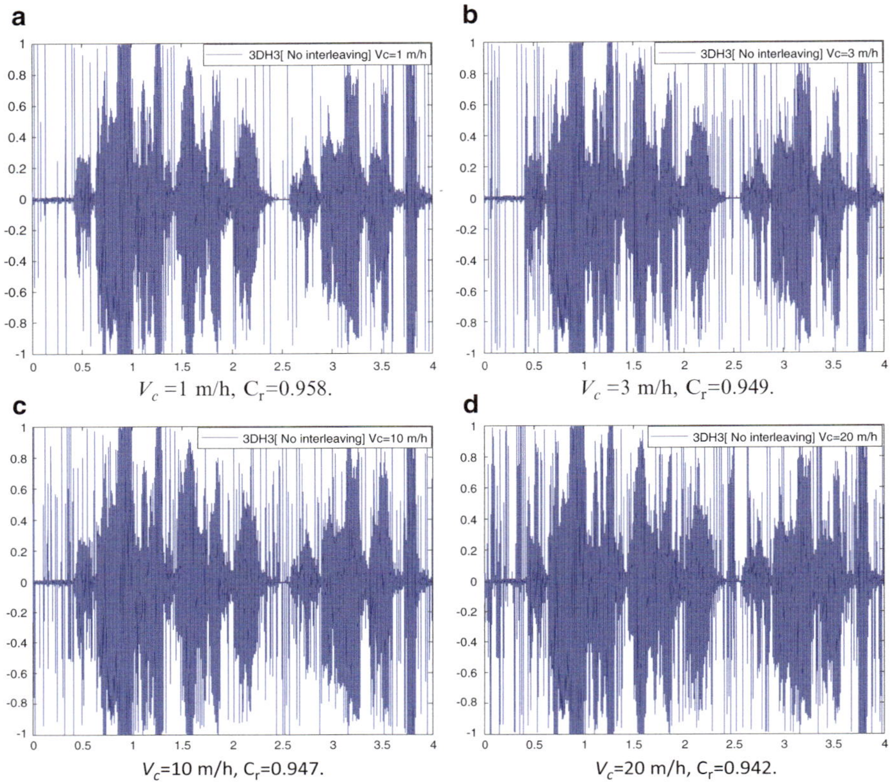

Fig. 6.21 Audio signal transmission using $3DH_3$ packets over a correlated Rayleigh fading channel at SNR = 15 dB

a power spectrum, with the variation of the signal power in the y axis and the sampling time (or the sample number) on the x axis [43].

In Fig. 6.21, it shows the effect of different velocity of the mobility on the received waveform of audio signal using Jakes model [11]. This example is devoted because of the widely using of the sound sensor in the WSN application such as the earthquake sensors or other fields [44].

This section is devoted for studying the effects of the mobility of the wireless ad hoc network on the transmitted packets and the received audio signals. As shown also, in Fig. 6.22, the different velocity of the mobility effect is cleared. Figure 6.23 shows the original audio signal waveform. In this computer experiment, the v is 1 up-to 20 miles/h. Figures 6.24, 6.25, 6.26, and 6.27 give the throughput, Cr, MSE, and NLF variation with the variety of the channel SNR with the shortest EDR uncoded $2DH_1$ packets, respectively. It is clear that with increasing the Vc of the terminal, the NLF is increased and the correlation between the original audio signal and the received signal is decreased. As shown in Fig. 6.24, the throughput is

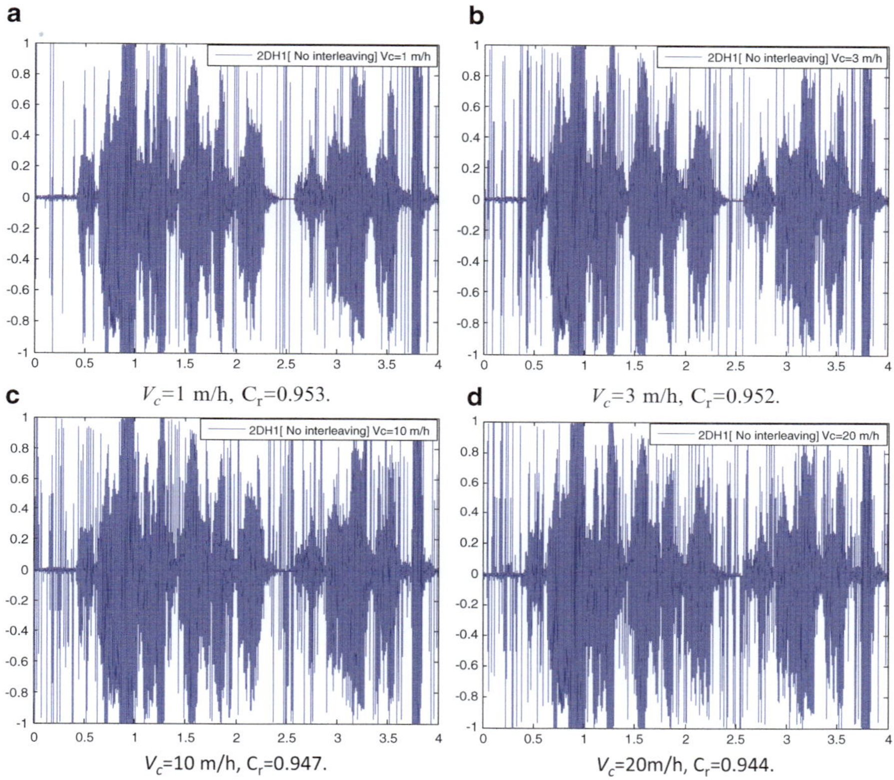

V_c=1 m/h, C_r=0.953.

V_c=3 m/h, C_r=0.952.

V_c=10 m/h, C_r=0.947.

V_c=20m/h, C_r=0.944.

Fig. 6.22 Audio signal transmission using $2DH_1$ packets over a correlated Rayleigh fading channel at SNR = 15 dB with different V_c

degraded with the velocity increase. Figure 6.25 gives the Cr variation with the different velocities. So, these results indicate the mobility effects on the different metrics of performance evaluation.

Also, the velocity bad effect on the performance of mobile wireless networks in general, the results of the MSE with variety of channel SNR is given in Fig. 6.26.

In the same manner the Mean Square Error of the transmitted audio signal over the mobile channel with the velocity variation and the variety of channel SNR is given in Fig. 6.25.

All these figures are investigated using the Jakes model to evaluate the velocity changes of the node in the mobile network. The using of audio signal here is related to the using of audio signal sensing using "microphone" in many applications of WSN such as natural phenomenal (earthquake for example).

The last metric is devoted as shown in Fig. 6.27 is the Number of Lost Frame (NLF). It gives the NLF with the variety of channel SNR with different velocity.

Fig. 6.23 Original waveform of the experiments

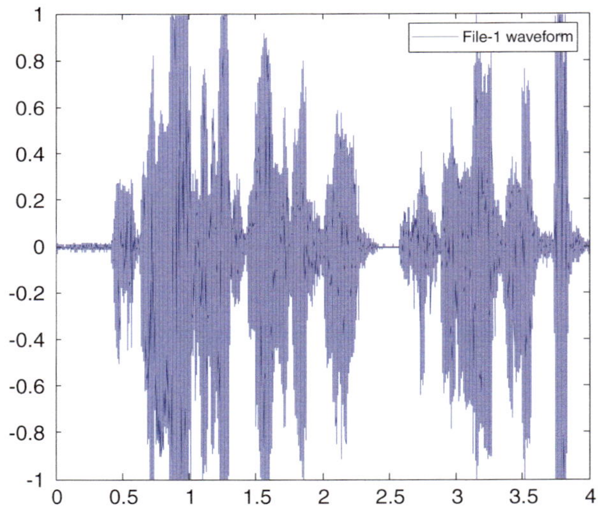

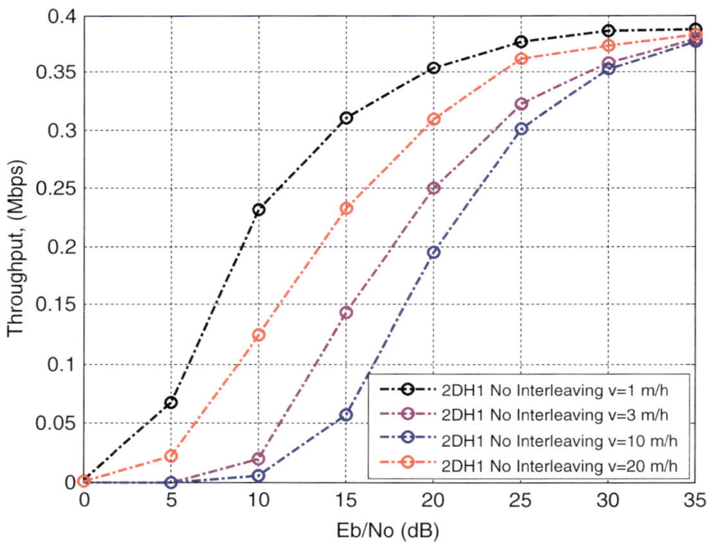

Fig. 6.24 Throughput of audio signal vs. SNR over correlated fading channel with different $V_c = 1$ up to 20 m/h

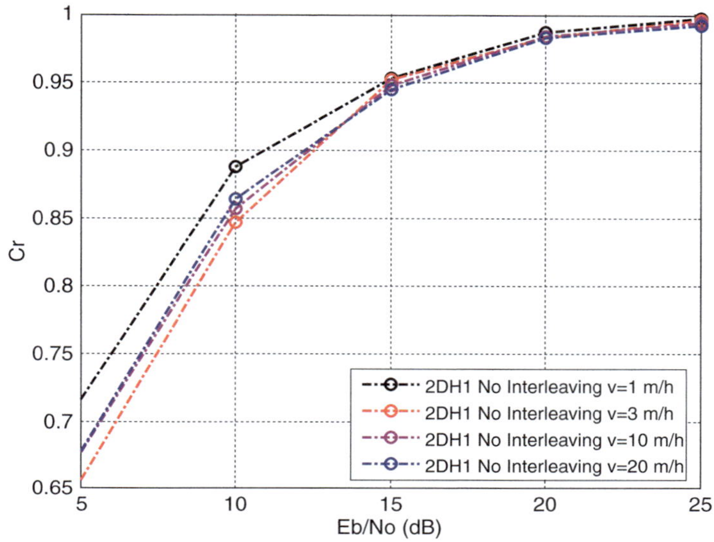

Fig. 6.25 Cr of audio signal vs. SNR for $2DH_1$ packets over correlated fading channel with V_c00A0 = 1 up to 20 m/h

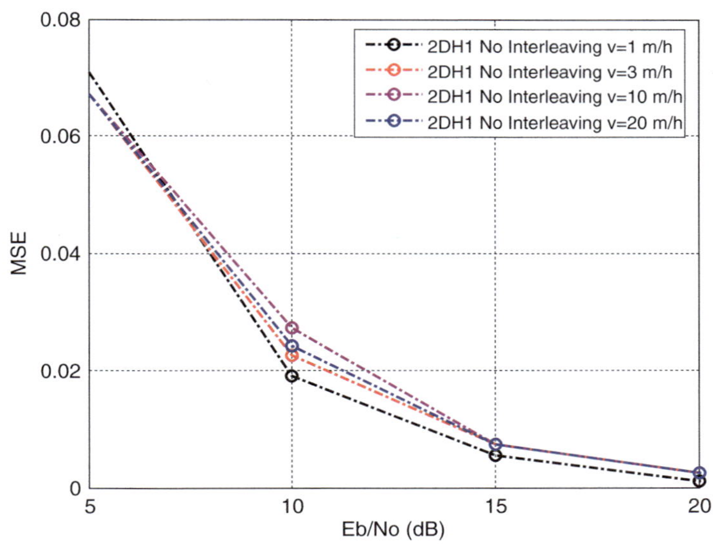

Fig. 6.26 MSE of audio signal vs. SNR for $2DH_1$ packets over correlated fading channel with different velocity

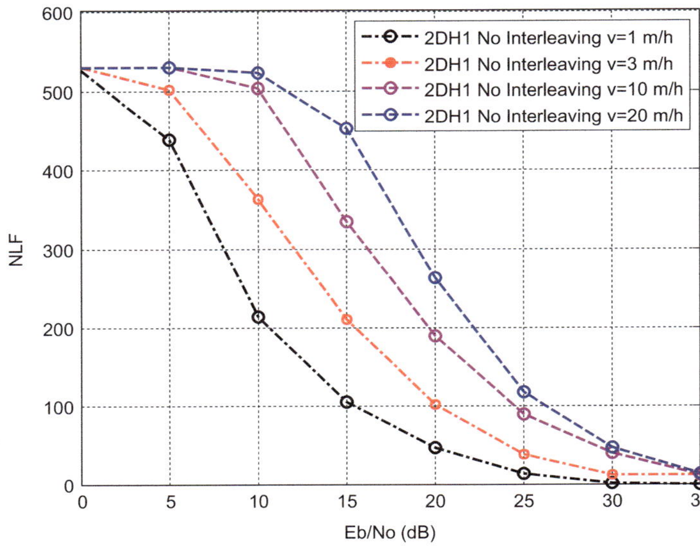

Fig. 6.27 NLF of audio signal vs. SNR over correlated fading channel with different V_c

****

MATLAB Code for Jakes Model which is utilized for simulation the mobility in these results simulation.

*************************_************************************

Matlab code of Jakes model:

```
###################################-############
function [fading]=jack_fading(L_total);
%generate rayleigh fading using Jack model.
%L_total: Numbers of elements require to generate.
%fading: an array of random variables whose elements are
%Rayleigh distributed in the interval defined by t.
%the following parameters are used to debug.
%L_total=9;
%L_total=2847;
%randomly choice observe start point.
o_start=rand(1,1)*10;
alpha=0; N0=8; N=(2*N0+1)*2;
sample=L_total; %how many bits needed
n=[1:N0]';
beta_n=n*pi/(N0+1);
%a car's velocity. default=60mile/hour.
v_c=10*1609/3600;
%air velocity
```

```
c=3*10^8;
%carrier frequency. default=2GHz
fc=2.46*10^9;
%doppler frequency.
omega_m=2*pi*v_c*fc/c;
omega_n=omega_m*cos(2*pi*n/N);
%sample interval;
delta_t=1/10000;
%t is observe time.
t=[o_start+delta_t:delta_t:o_start+delta_t*sample];

for k=1:N0
   temp_x_c(k,:)=cos(beta_n(k))*cos(omega_n(k)*t);
   temp_x_s(k,:)=sin(beta_n(k))*cos(omega_n(k)*t);
end
x_c=(sum(temp_x_c)*2+sqrt(2)*cos(alpha)*cos(omega_m*t))/sqrt
(N0);
x_s=(sum(temp_x_s)*2+sqrt(2)*sin(alpha)*cos(omega_m*t))/sqrt(N0
+1);

fading=sqrt(x_c.^2+x_s.^2)/sqrt(2);%sum along the row.

%plot to see the figure.
%x=1:1:L_total;
%y=jack_fading;
%plot(x,y,'-');
%%plot the title and label
%title('jack model');
%ylabel('Rayleigh fading Amplitude');
%xlabel('Sample time');
####################################_##############
```

6.9 Mobility Effects

As shown in the previous section, the mobility has bad effects on the performance [45, 46]. In this paper the mobility effect is investigated on the image transmission over WSN ZigBee based. It presents the following assumption, the mobile ZigBee device velocity is 10 miles/h, and the carrier frequency is 2.46 GHz. The Doppler spread is 36.6 Hz. The results are given using the standard Matlab the Cameraman image used in the experiments. It is the Matlab image and it format is Tag Image File format (TIF).

As shown in the paper results, the mobility has bad effects on the PSNR values of the received images as shown in Fig. 6.28. These experiments studied different velocity of the mobile terminal with different SNR of the communication channels.

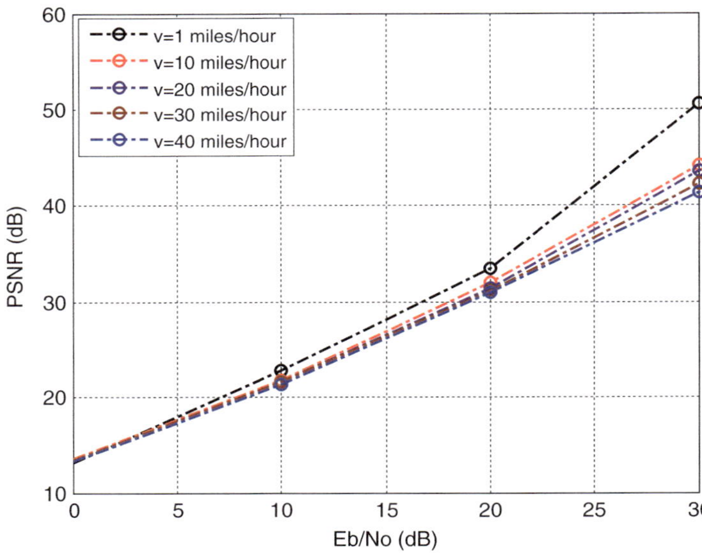

Fig. 6.28 PSNR vs. SNR for the received Cameraman image over the ZigBee standard with the different *Vc*

Also, the Number of Lost Packets (NLP) is used as a metric for measuring the mobility effects. Figure 6.29 gives the NLP variation with the channel SNR. It is clear that with increasing the V_c of the terminal, the NLP is increased. With the same expected manner, Fig. 6.30 shows the BER variation with variety SNR and different velocity [47].

In the following Fig. 6.31, received image samples have been presented. As shown in this sample the velocity of the mobility has bad effects on the received image. The PSNR of every image sample is mentioned below the figure. The scenario is repeated with two SNR of the channel also, the Jakes model is utilized in the simulation.

As shown in the previous figures, the mobility has bad effects on the PSNR values of the received images. These experiments studied different velocity of the mobile terminal with different SNR of the communication channels.

The previous scenario is repeated with another SNR of the channel. Also, the PSNR value of the received image samples is used as a image metric quality. This results is devoted to ensure the mobility effects, also, the positive effects on the mobility of sensing node more than the bad effects or not. Figure 6.32 shows the received image samples with the channel SNR = 20 dB and the velocity of the mobility variations.

The mobility feature has bad effect on the WSN performance. So, the mobile WSN presents flexible sensing station, its mobility may reduce the false alarm, but the mobile WSN performs worse than the fixed one. Therefore, the mobility can be feature, it is provided to determined number of sensing node station. These mobile

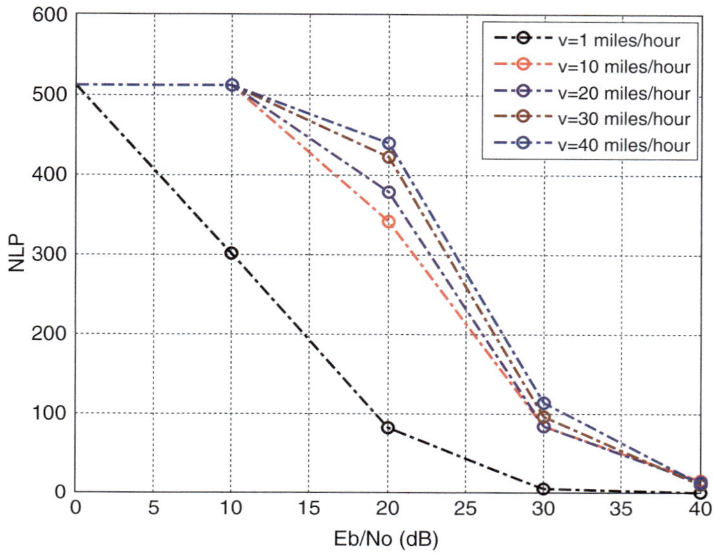

Fig. 6.29 NLP vs. SNR for the received Cameraman image over the ZigBee standard with the different *Vc*

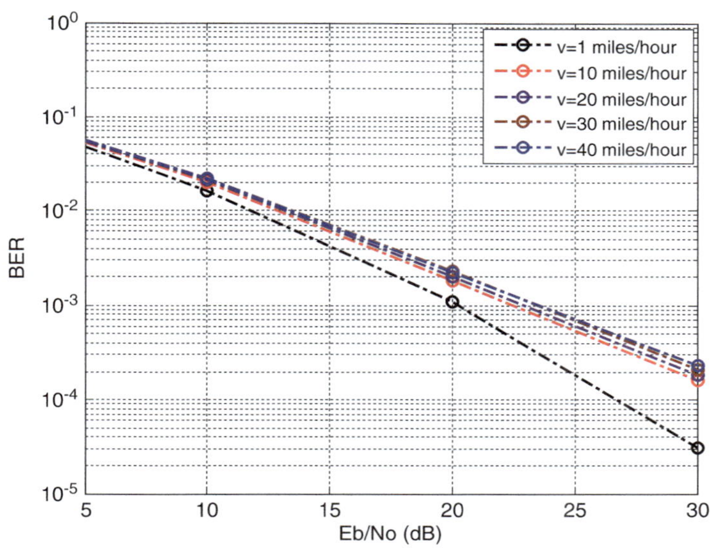

Fig. 6.30 BER vs. SNR for the received Cameraman image over the ZigBee standard with the different *Vc*

Fig. 6.31 Received Cameraman image over a correlated fading channel at SNR = 10 dB with (**a**) PSNR = 22.7 dB, (**b**) PSNR = 21.7 dB, (**c**) PSNR = 21.6 dB, (**d**) PSNR = 21.3 dB, and (**e**) PSNR = 21.2 dB

Fig. 6.32 Received Cameraman image over a correlated fading channel at SNR = 20 dB with (**a**) PSNR = 33.47 dB, (**b**) PSNR = 31.97 dB, (**c**) PSNR = 31.38 dB, (**d**) PSNR = 31.1 dB, and (**e**) PSNR = 31.0 dB

nodes can be called "Emergency Sensing Nodes" (ESN). The mobility effects on the error performance of the ZigBee network through BER variation with the channel SNR is shown in Fig. 6.30. It is clear that with increasing the *Vc* of the terminal, the error performance of the system is degraded [48].

6.10 Cooperative Algorithm of WSN

The WSN-ZigBee based performance is presented with the standard and some proposed scenarios simulation in this chapter. The mobility ability of WSN nodes is also considered in the simulation and modeling code. The chapter presents cooperative algorithm between WSN nodes through the proposed interleaving technique. Also, it focuses on the historical notes about the interleaving techniques art. The traditional techniques are just reordering data tools. The presented algorithm is based on utilizing the most common data randomizing tools which own a secret key. This chapter introduces and studies a vital issue in wireless communications, which is the transmission of images over wireless networks. IEEE ZigBee 802.16.4 is a short-range communication standard that could be used for small distance multimedia transmissions. In fact, the ZigBee network is a Wireless Personal Area Network (WPAN), which needs a strong interleaving mechanism for protection against error bursts. This chapter presents a novel chaotic interleaving scheme for this purpose. This scheme depends on the chaotic Baker map. A comparison study between the proposed chaotic interleaving scheme and the traditional block and convolutional interleaving schemes for image transmission over a correlated fading channel is presented. The simulation results show the superiority of the proposed chaotic interleaving scheme over the traditional schemes.

The meaning of WSN, it is wireless network contains many numbers of sensing node. WSN means many sensing station (SS) "denoted by nodes", these nodes cover and service specific volume for monitoring and sense any changing happening. By default there is cooperative working between the nodes. This cooperation is not announced, but it is understood through the sensing concept. The number of nodes must e enough the volume area, every sensor can cover specific area called "the coverage area", it part from the whole service area "volume". So, the service area contains number of sensor coverage area.

With the service area as shown in the Fig. 6.33 the WSN consists of spatially distributed autonomous sensors, every one of them has specific converge area, it should be overlapped with the adjacent converge area of another nodes, where in the reality design and deploying the SN, the spaces between the coverage areas mush be not found. The sensors cooperatively monitor physical or environmental conditions. So, it can be considered that the WSN is provided by the cooperatively working network concept. As shown in Fig. 6.33, Service area may be contained uncovered areas are free areas. These areas are not covered by any detectors.

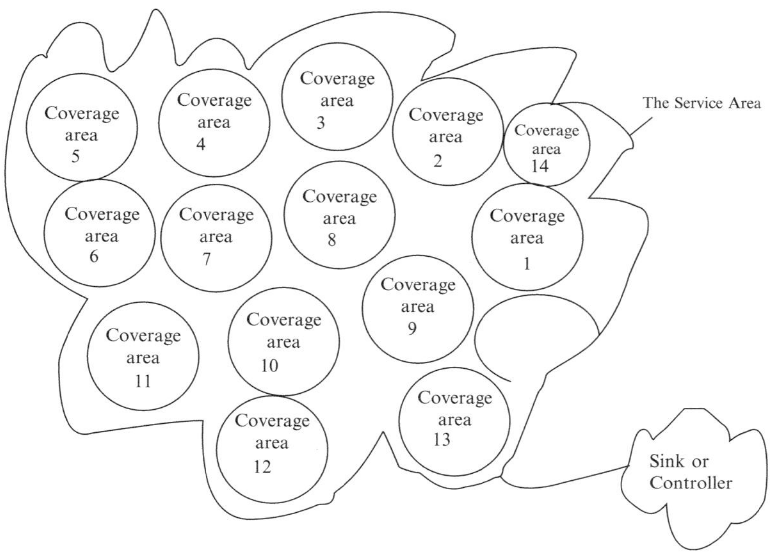

Fig. 6.33 Service area and coverage sensor area relation

The proposed view of cooperative WSN nodes, the cooperative algorithm is presented to enhance the cooperation between the sensing stations of the specific area. The algorithm enables the nodes talk together and request sharing denying or confirming the capture data. The algorithm can be handled by the sink or the head on cluster node. Also, it may be self organizing decision by the node.

6.10.1 EMSN

EMSN refers to "Emergency Mobile Sensing Node". The mobile nodes may be operated in the hot areas and high sensitive area in the service area of the WSN. The idea of ESN has another benefit, it provides continuous compensation for any lack of a number monitoring points, especially in areas such as widening vast forests, beaches, and etc.

6.10.2 WSN Ad-Hoc Networks

Figure 6.34 gives a simple imagine for the spatially changes of the coverage area with the mobility of the sensing node. Every sensor has limited coverage area, so, the mobility of sensor station means spatial changing of this coverage area also, may decrease its sensing capability. The mobile sensing node should be additional

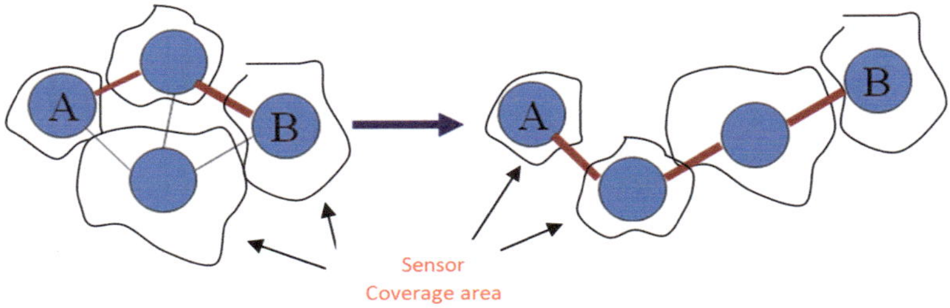

Fig. 6.34 Spatial deployment of SN with the Mobile nodes

nodes or as called previously emergency sensing node. It serves as a Rapid Reaction Forces (RRF) with quick tasks and less time working as possible.

The cooperation between the sensing nodes through unique secret key every number of nodes group (sink server). With employing the previous technique of the interleaving chaotic based, it is very easy to share the same secret for define adjacent sensor.

In brief, it should be noted that:

- Wireless Sensor Networks are a special category of ad hoc networks that are used to provide a wireless communication infrastructure among the sensors deployed in a specific application domain.
- A sensor network is a collection of a large number of sensor nodes that are deployed in a particular region.
- Distinct properties of wireless sensor networks:
  - Mobility of nodes is not needed in all cases in wireless sensor networks without cooperative algorithm between the nodes.
  - The size of the network is much larger than that in a typical ad hoc wireless network.
  - The density of nodes in a sensor network varies with the domain of application.

6.10.3 Mobility Features

The WSN contains a group of mobile sensing station through the sink decision, the network called "Mobile Ad-Hoc Sensor Network". The mobility ability of sensing or monitor nodes is managed by the cooperation algorithm, this algorithm is controlled and operated by the node itself, the WSN here is called intelligent Mobile Ad-Hoc sensor network. The benefits of the cooperative algorithm are several; it decreases the need or huge numbers of sensing station to cover specific area, also, the capture event by nearest sensor can be confirmed by another mobile sensing station. The calling between the sensors can be available.

Sensor nodes may change their location after initial deployment. Mobility can result from environmental influences such as wind or water, sensor nodes may be attached to or carried by mobile entities, and sensor nodes may possess automotive capabilities. In other words, mobility may be either an incidental side effect, or it may be a desired property of the system (e.g., to move nodes to interesting physical locations), in which case mobility may be either active (i.e., automotive) or passive (e.g., attached to a moving object not under the control of the sensor node). Mobility may apply to all nodes within a network or only to subsets of nodes. The degree of mobility may also vary from occasional movement with long periods of immobility in between, to constant travel.

Mobility has a large impact on the expected degree of network dynamics and hence influences the design of networking protocols and distributed algorithms. The actual speed of movement may also have an impact, for example on the amount of time during which nodes stay within communication range of each other. Also, as shown in the previous scenarios which has been investigated the movement speed on the audio and image transmission, it reveals that the accepted performance of mobile WSN needs more powerful techniques, that costs the system additional complexity and additional time plus the power consumption.

"Although, all the mobility constraints, it is still interesting option. It is still an option that can increase the amount of suspense surrounding this kind of networking and provides additional importance, as well as increases the areas of uses."

References

1. Singh S, Raghavendra CS (1998) PAMAS: Power Aware Multi-access Protocol with Signaling for ad hoc networks. ACM Comput Commun Rev 28(3):5–26
2. Rappaport TS (1996) Wireless communications: principles and practice. Prentice Hall Communications Engineering and Emerging Technologies Series, Upper Saddle River
3. Perez-Vega C, Garcia JL (1997) A simple approach to a statistical path loss model for indoor communications. In: 27th European microwave conference and exhibition, Jurusalem, Israel, pp 617–623
4. Hansen J, Leuthold PE (2003) The mean received power in ad hoc networks and its dependence on geometrical quantities. IEEE Trans Ant Prop 51(9):2413–2419
5. Howard SL, Schlegel C, Iniewski K (2006) Error control coding in low-power wireless sensor networks: when is ECC energy-efficient? EURASIP J Wirel Commun Netw 2006:074812. doi:10.1155/WCN/2006/74812
6. Green DB, Obaidat AS (2002) An accurate line of sight propagation performance model for ad-hoc 802.11 wireless LAN (WLAN) devices. IEEE Intl Conf Commun (ICC) 5:3424–3428
7. Garello R, Montorsi G, Benedetto S, Cancellieri G (2001) Interleaver properties and their applications to the trellis complexity analysis of turbo codes. IEEE Trans Commun 1:793–807
8. Abd El-Samie FE, Hassan ES, Zhu X, El-Khamy SE, Dessouky MI, El-Dolil SA (2010) A chaotic interleaving scheme for the continuous phase modulation based single-carrier frequency-domain equalization system. Wirel Pers Commun 62:183–199, Springer Science+-Business Media, LLC

9. Hagenauer J, Papke L (1996) Iterative decoding of binary block and convolutional codes. IEEE Trans Inf Theory 42(2):429–445
10. Gibson JD (1993) Principles of digital and analog communications, 2nd edn. University of California, Santa Barbara. ISBN 0-02-341860-5
11. El-Bendary MAM, Abou-El-azm AE, El-Fishawy NA, Shawki F, El-Tokhy MAR, Abd El-Samie FE, Kazemian HB (2012) JPEG image transmission over mobile network with an efficient channel coding and interleaving. Int J Electron 99(11):1497–1518. doi:10.1080/00207217.2012.680786
12. Kaplan JL, Yorke JA (1979) Lecture Notes Math 730:204
13. Dong Y, Liu L, Zhu C, Wang Y (2010) Image encryption algorithm based on chaotic mapping. In: 3rd IEEE international conference on computer science and information technology, ICCSIT, pp 289–291
14. Salleh M, Ibrahim S, Isnin IF (2003) Enhanced chaotic image encryption algorithm based on Baker's map. IEEE Conf Circuit Syst 2:508–511
15. Yuan DF, Li ZW, Sui A, Luo J (2000) Performance of interleaved (2,1,7) convolutional codes in mobile image communication system. In: Proceedings of the IEEE wireless communications and networking conference (WCNC '00), Chicago, vol 2, pp 634–637
16. Chan F, Haccoun D (1997) Adaptive viterbi decoding of convolutional codes over memoryless channels. IEEE Trans Commun 45(11):1389–1400
17. Benvenuto N, Bettella L, Marchesani R (1998) Performance of the viterbi algorithm for interleaved convolutional codes. IEEE Trans Veh Technol 47(3):919–923
18. Kong JJ, Parhi KK (2003) Interleaved convolutional code and its viterbi decoder architecture. EURASIP J Appl Signal Process 13:1328–1334
19. Elkhazin A, Plataniotis K, Pasupathy S (2010) Irregular convolutional codes in multiantenna bit-interleaved coded modulation under iterative detection and decoding. IEEE Trans Veh Technol 59(7):3332–3341
20. Callegari S, Rovatti R (1999) Analog chaotic maps with sample-and-hold errors. IEICE Trans Fundam Electron Commun Comput Sci E82A(9):1754–1761
21. Wu Y, Yang G, Jin H, Noonan JP (2012) Image encryption using the two-dimensional logistic chaotic map. Electron Imaging 21(1):013014. doi:10.1117/1.JEI.21.1.013014
22. Golmie N, Van Dck RE, Soltanian A (2001) Interference of Bluetooth and IEEE 802.11: simulation modeling and performance evaluation. In: Proceedings ACM international workshop on modeling, analysis, and simulation of wireless and mobile systems, Italy
23. Chui TY, Thaler F, Scanlon WG (2002) A novel channel modeling technique for performance analysis of Bluetooth baseband packets. In: Proceedings of the IEEE ICC conference, New York
24. Hall EK, Wilson G (2001) Stream-oriented turbo codes. IEEE Trans Inf Theory 47(5):1813–1831
25. Haccounand D, Begin G (1989) High-rate punctured convolutional codes for viterbi and sequential decoding. IEEE Trans Commun 37(11):1113–1125
26. Proakis JG (2001) Digital communications. McGraw-Hill Series in Electrical and Computer Engineering, University of Michigan, Ann Arbor, New York
27. Voyatzis G, Pitas I (1998) Chaotic watermarks for embedding in the spatial digital image domain. Proc IEEE Int Conf Image Process 2:432–436
28. El-Bendary MAM, Abou-El-azm AE, El-Fishawy NA, Shawki F, Abd-ElSamie FE, El-Tokhy MAR, Kazemian HB (2012) Performance of the audio signals transmission over wireless networks with the channel interleaving considerations. EURASIP J Audio Speech Music Process 2012:4
29. Lu W, Tao H, Chung F (2005) Chaos-based spread spectrum robust watermarking in DWT domain. In: Fourth international conference on machine learning and cybernetics, Guangzhou, 18–21 Aug 2005
30. Herzberg H (2006) Multilevel turbo coding with short interleavers. IEEE J Sel Areas Commun 16:303–309

31. Kaiser M, Fong W, Sikora M (2009) A comparison of decoding latency for block and convolutional codes. In: Proceeding, ISCTA'09, Ambleside
32. Lin L, Cheng RS (1997) Improvements in SOVA-based decoding for turbo codes. IEEE Global Telecommun Conf (Globecom) 2:644–648
33. Jakes WC (1975) Microwave mobile communications. Wiley, New York. ISBN 0-471-43720-4
34. Forney GD (1971) Burst-correcting codes for the classic bursty channel. IEEE Trans Commun COM-19:772–781
35. Vafi S, Wysocki TA (2006) Application of convolutional interleavers in turbo codes with unequal error protection. JTIT, J Telecommun Inf Technol 1:17–23
36. Vafi S, Wysocki T (2005) Performance of convolutional interleavers with different spacing parameters in turbo codes. In: Proceedings of the 6th Australian workshop on communications theory, Brisbane, Qld pp 8–12
37. Shiyamala S, Rajamani V (2010) A novel area efficient folded modified convolutional interleaving architecture for MAP decoder. Int J Comput Appl 9(9):975–8887
38. Zhang H, Wang L, Yuan Q, Wang H, Yu J (2004) A chaotic interleaver used in turbo codes. Int Conf Commun Circuit Syst (ICCCAS) 1:27–29
39. Xuelan Z, Weiyan L, Guangzeng F (2010) Applying chaotic maps to interleaving scheme design in BICM-ID. Chin J Electron 19(3)
40. Luby M, Mitzenmacher M, Shokrollahi MA, Spielman DA (2001) Efficient erasure correcting codes. IEEE Trans Inf Theory 47(2):569–584
41. Počta P (2011) The combined effect of signal strength and background traffic load on speech quality in IEEE 802.11 WLAN. Radioengineering 20(1):174–178
42. Mohamed MAM, Abou El-Azm A, El-Fishwy N, El-Tokhy MAR, Abd El-Samie FE (2008) Optimization of Bluetooth packet format for efficient performance. Progr Electromagn Res M 1:101–110
43. Aldrich J (1995) Correlations genuine and spurious in Pearson and Yule. Stat Sci 10:364–376. http://www.jstor.org/stable/2246135
44. Galli S, Famolari D, Kodama T (2004) Bluetooth: channel coding considerations. In: IEEE vehicular technology conference VTC pp 2605–2609
45. El-Bendary MAM, Abou-El-azm AE, El-Fishawy NA, Shawki F, El-Tokhy M, Abd El-Samie FE, Kazemian HB (2011) Enhancing the image transmission over wireless networks through a novel interleaver. KSII Trans Internet Inf Syst 5(9):1528–1543
46. El-Bendary MAM, El-Tokhy MAR (2013) An efficient burst error combating for image transmission over mobile WPANs. World Acad Sci Eng Technol Int J Electric Electron Sci Eng 7(5)
47. Ozarow L, Shamai S, Wyner AD (1994) Information theoretic considerations for cellar mobile radio. IEEE Trans Veh Tech 43:359–378
48. Pekhteryev G, Sahinoglu Z, Orlik P, Bhatti G (2005) Image transmission over IEEE 802.15.4 and ZigBee networks. In: IEEE ISCAS, Kobe

Chapter 7
Appendix

7.1 Enhancing the Transmission over WSN ZigBee Based Scenario

7.1.1 Simulation Scenario 1

Many researches present the WSN implementation using the ZigBee technology. In the presented scenario 1, there several sub-scenarios are studied. The mobility effects, the different data randomizing improving degree, and the effectiveness of the convolutional code constraint length on the image transmission over the mobile channel. The scenario tried to find efficient technique alternative to the complex Convolutional decoder.

7.1.1.1 Scenario Definition and Goal

This scenario presents different scenarios for the efficient transmission of images over wireless networks. The scenario focuses on the use of the IEEE 802.16.4 ZigBee, which is a Wireless Personal Area Network (WPAN), for applying the proposed scenarios. The heart of these scenarios is a novel chaotic interleaving scheme, which can tolerate error bursts. The scenario studies the performance of the proposed interleaver with convolutional codes having different constraint lengths (K). A comparison study between the traditional block interleaving scheme and the proposed chaotic interleaving scheme for image transmission over a correlated fading channel is presented. The simulation results show the superiority of the proposed chaotic interleaving scheme. The results also prove that the chaotic interleaver on a packet-by-packet basis gives a high quality image with ($K = 3$) and eliminates the need for a complex encoder having $K = 7$.

M.A.M. El-Bendary, *Developing Security Tools of WSN and WBAN Networks Applications*, Lecture Notes in Electrical Engineering 316, DOI 10.1007/978-4-431-55069-3_7

7.1.2 Scenario Motivations

With the increase in utilization of wireless networks, there are two important factors that deserve consideration; power efficiency and throughput. Over bad communication channels, different Forward Error Correction (FEC) schemes are used [1]. In this scenario, the convolutional coding is used as an FEC technique, because it can significantly improve the performance of communication systems. The Viterbi algorithm is one of the most widely used decoding algorithms for convolutional codes. It is optimal, but its complexity in both the number of computations and memory requirements is exponentially increased with the constraint length K of the code. So, practically, the constraint length K of convolutional codes is limited to $K < 9$. Hence, when codes with longer constraint lengths are required in order to achieve a lower error probability, decoding algorithms whose complexity does not depend on the constraint length become especially attractive [2, 3].

In fact, the environment of mobile communications suffers from long burst errors. In the fast fading case, these burst errors are due to the multipath effect, and they reduce the quality of communication. An error control technique in mobile communication is not enough with long burst errors [4]. In [5], the effect of interleaved convolutional coding with code rate 1/2 and $K = 7$ on image transmission over a wireless channel was studied. The adaptive Viterbi decoding for interleaved convolutional coding was presented in [6] and [7], but with little improvement. In fact, there is a need for an effective error spreading tool to enhance the quality of communication.

This scenario presents a novel technique to achieve a lower error probability with a small constraint length of the convolutional encoder. This technique depends on the data interleaving or randomization. So, long burst errors can be separated into single random or short burst errors. Computer simulations have been carried out on the proposed chaotic interleaver for the ZigBee network assuming the Jakes channel model. The simulation results have revealed that the proposed technique is efficient for image communication over the ZigBee network [8].

7.1.3 Scenario Material

For evaluating and testing the effectiveness and the performance the scenario, the ZigBee network is chosen.

The short-range wireless networks such as the Bluetooth and ZigBee are widely used in essential applications such as health care, medical applications, home automation, and environmental applications [9, 10]. The ZigBee network is a Low-Rate WPAN (LR-WPAN) that is used for short-range, low-power, and low-cost data communication. The structure of the ZigBee packet is shown in Fig. 7.1. The header contains three fields; a preamble of 32 bits for synchronization, a packet delimiter of 8 bits, and a physical header of 8 bits. The Physical Service

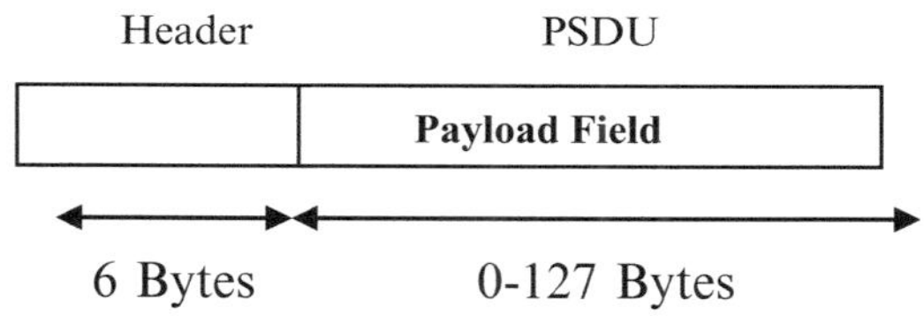

Fig. 7.1 ZigBee packet format

Data Unit (PSDU) field contains a payload of 0–127 bytes length. The ZigBee network uses an error detection/retransmission technique. To ensure successful reception of data, an acknowledged frame delivery protocol is supported to increase transfer reliability [11].

Low-power consumption in ZigBee networks can be achieved by allowing a device to sleep, which means waking into active mode for brief periods. Enabling such low-duty-cycle operation is at the heart of the ZigBee standard. ZigBee is built on top of the IEEE 802.16.4 standard. It offers the additional functionality to implement mesh networking rather than point-to-point networking found in most Bluetooth and Wi-Fi applications. The ZigBee specification document is short, allowing a small and simple stack, in contrast to the other wireless standards such as Bluetooth [12].

The IEEE 802.16.4 standard is intended to conform to established regulations in Europe, Japan, Canada, and the United States. It defines two physical (PHY) layers; the 2.4 GHz and 868/915 MHz band PHY layers. Although the PHY layer chosen depends on local regulations and user preference, only the higher data rate, worldwide, unlicensed 2.4 GHz Industrial, Scientific, and Medical (ISM) frequency band is considered. A group of 16 channels is available in the 2.4 GHz band, numbered from 11 to 26, each with a bandwidth of 2 MHz, and a channel separation of 5 MHz. The channel mapping frequencies are given in Table 7.1. The LR-WPAN output powers are around 0 dBm. It typically operates within a 50-m range. The transmit scheme used is the Direct Sequence Spread Spectrum (DSSS) [13–15].

7.1.4 Scenario Preparations

The transmission of multimedia over unreliable data links has become a topic of paramount importance. This type of transmission must reconcile the high data rates involved in multimedia contents and the noisy nature of the channels, be it wireless or mobile. In this scenario, there are different proposed scenarios to improve the transmission of images over the ZigBee network through interleaving. We study the feasibility of data interleaving prior to transmission over ZigBee networks. The scenario presents a new chaotic interleaver and compares it to the traditional block and convolutional interleavers.

In a general convolutional code, the input information sequence contains $k \times L$ bits, where k is the number of parallel information bits at a time interval, and L is the number of time intervals. This results in $m+L$ stages in the Trellis diagram,

Table 7.1 IEEE 802.16.4 frequency bands and data rates

PHY (MHz)	Freq. band (MHz)	Mod.	Channels	Bit rate (kbps)
868/915	868–868.6	BPSK	1	20
	902–928	BPSK	10	40
2,450	2,400–2,483.5	O-QPSK	16	250

where m is the number of shift registers in the encoder and $\mathrm{K} = m + 1$. There are exactly $2^k \times L$ distinct paths in the Trellis diagram. As a result, the Maximum-Likelihood (ML) sequence would have a computational complexity of $\mathrm{O}(2^k \times L)$. The Viterbi algorithm reduces complexity by performing the ML search on a stage at a time in the Trellis diagram at each node. The number of nodes per stage in the Trellis diagram is 2^m. Therefore, the complexity of Viterbi calculations is of $\mathrm{O}((2^k)$ (2^m) $(m + L))$ [16]. This significantly reduces the number of calculations required to implement the ML decoding, because the number of time intervals L is now smaller [17, 18]. With the increase in m and k, the complexity is increased exponentially. So, the scenario tries to reduce the computational complexity through the proposed chaotic interleaver by eliminating the need of convolutional codes with large constraint lengths.

The computer simulations are based on binary non-recursive convolutional coding. The convolutional encoder uses a constituent encoder with K = 3, 5, or 7, code rate = 1/2 and generator polynomials, G = (5,7), (23,35), or (133,171) in octal [18]. The scenario uses the proposed chaotic interleaving technique to improve the capabilities of the convolutional codes with short constraint lengths.

7.1.4.1 Burst Error with Traditional Interleaver

The block interleaving can be used for image transmission with the ZigBee network. After converting the image into a binary sequence, this sequence is rearranged into a matrix in a row-by-row manner, and then read from the matrix in a column-by-column manner. Now take a look at how the block interleaving mechanism can correct error bursts. Assume an error burst affecting four consecutive bits (1-D error burst) as shown in Fig. 7.2b with shades. After de-interleaving as shown in Fig. 7.2c, the error burst is effectively spread among four different rows, resulting in a small effect for the 1-D error burst. With a single-error correction capability, it is obvious that no decoding error will result from the presence of such 1-D error burst. This simple example demonstrates the effectiveness of the block interleaving mechanism in combating 1-D error bursts. Let us examine the performance of the block interleaving mechanism, when a 2-D (2×2) error burst occurs [19], as shown in Fig. 7.3b with shades. Fig. 7.3c indicates that the 2×2 error burst has not been spread, effectively, so that there are adjacent bits in error in the first and second rows. As a result, this error burst cannot be corrected using a single-error correction mechanism. That is, the block interleaving mechanism cannot combat the 2×2 error bursts [20].

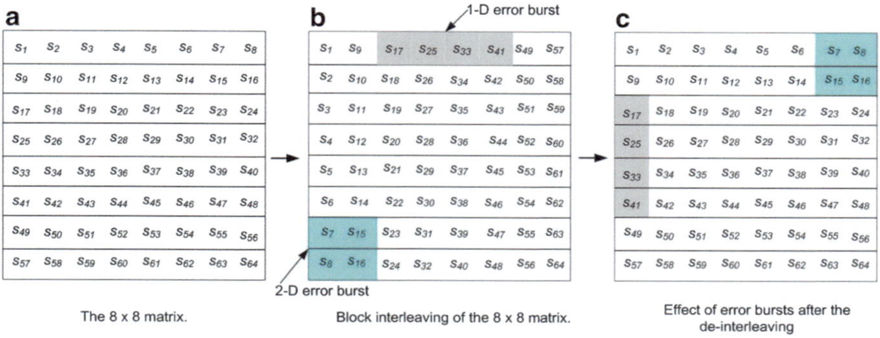

Fig. 7.2 Block interleaver example (8 × 8 square matrix)

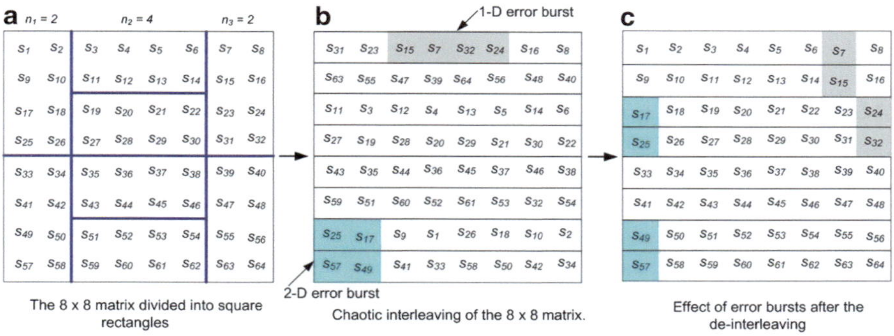

Fig. 7.3 Chaotic interleaver example (8 × 8 square matrix)

7.1.4.2 Burst Error with the Convolutional Interleaver

The convolutional interleaver is constructed by T parallel branches. Each line contains a shift register with a predefined length [21]. The input data is fed into the branches of the interleaver and the output data is taken from the outputs of these branches. In the computer simulations, the length of the interleaver input is 1,024 bits, which is the length of the whole payload in ZigBee packets [22].

7.1.4.3 Burst Error with the Chaotic Interleaver

The chaotic interleaver idea is presented using the logistic map in [23, 24]. This map is defined as a typical example of a 1-D chaotic map. In [23], it is applied with the turbo code. In [24], a logistic interleaver is employed in a Bit-Interleaved Coded Modulation (BICM) technique. In this work, the proposed chaotic interleaver is based on a 2-D Baker map, which is a powerful data randomization tool. This map is widely used as an encryption tool [24, 25], and it will be developed in this scenario as an interleaving tool.

From Fig. 7.2, it is clear that the block interleaver is not efficient with 2-D error bursts. As a result, there is a need for an advanced interleaver for this task. The 2-D chaotic Baker map in its discretized version is a good candidate for this purpose. After rearrangement of bits into a 2-D format, the chaotic Baker map is used to randomize the bits. Let $B(n_1, \ldots, n_k)$, denote the discretized map, where the vector, $[n_1, \ldots, n_k]$, represents the secret key, S_{key}. Defining N as the number of data items in one row, the secret key is chosen such that each integer n_i divides N, and $n_1 + \ldots + n_k = N$.

Figure 7.3 shows an example for chaotic interleaving of an 8×8 square matrix (i.e. $N = 8$). The secret key, $S_{key} = [n_1, n_2, n_3] = [2, 4, 2]$. Note that the chaotic interleaving mechanism has a better treatment to both 1-D and 2-D error bursts than the block interleaving mechanism. Errors are better distributed to bits after de-interleaving in the proposed chaotic interleaving scheme. As a result, a higher Peak Signal-to-Noise Ratio (PSNR) of received images can be achieved with this proposed mechanism. Moreover, it adds a degree of security to the communication system. At the receiver of the ZigBee system, a chaotic de-interleaving step is performed.

7.1.5 Scenario Experiments

In this section, the computer simulation results are presented. A correlated Rayleigh fading channel is used. The channel model utilized is the Jakes model [26, 27]. The assumed mobile ZigBee device velocity is 10 miles/h, and the carrier frequency is 2.46 GHz. The Doppler spread is 36.6 Hz. Figure 7.4 gives the original Cameraman image used in the experiments. The computer simulation section contains three categories of experimental simulations.

7.1.5.1 Experiments Group 1

In this section, different scenarios are employed for the image transmission over the ZigBee network. The image binary sequence to be transmitted is fragmented into packets. The PSNR of the received images is used as an evaluation metric in this scenario.

In the first experiment, the Cameraman image is transmitted over a correlated-fading channel with Signal-to-Noise Ratio (SNR) = 10 dB. Different scenarios of no interleaving, block interleaving, convolutional interleaving and chaotic interleaving are considered for comparison. The results of this experiment are shown in Fig. 7.5. From these results, it is clear that the effects of all interleaving schemes is approximately equal at low SNR values.

Fig. 7.4 Original Cameraman image

Another experiment is repeated with SNR = 20 and 30 dB and the results are shown in Figs. 7.6 and 7.7. From these results, we notice that the chaotic interleaver outperforms the other interleavers at moderate and high SNRs.

For the comparison purpose, the variation of the PSNR of the received image, the number of lost packets or frames and the Bit Error Rate (BER) with the channel SNR are studied and the results are shown in Figs. 7.8 and 7.9. From these results, it is clear the chaotic interleaver enhancement begins at medium SNR values.

7.1.5.2 Experiments Group 2

This section is devoted to investigate the effect of the constraint length K of the convolutional code on the image transmission process. The Cameraman image is transmitted over a correlated fading channel having SNR = 10 dB using convolutional coding only with code rate = 1/2, and without interleaving. Constraint lengths of K = 3, 5, and 7 are investigated. Simulation results for this section are shown in Fig. 7.10. From these results, it is clear that the effect of the constraint length on the received image is marginal at SNR = 10 dB. The variations of the PSNR of the received image with the channel SNR for different constraint lengths are shown in Fig. 7.11. It is clear from that figure that a large constraint length gives better PSNR performance, especially at medium and high SNR values in the channel at the expense of an increased complexity.

Fig. 7.5 Received Cameraman image over a correlated fading channel at SNR = 10 dB with (**a**) PSNR = 21.3 dB, (**b**) PSNR = 21.4 dB, (**c**) PSNR = 21.1 dB, and (**d**) PSNR = 21.5 dB

7.1.5.3 Experiments Group 3

The proposed chaotic interleaver can be implemented on a packet-by-packet basis. Simulation experiments are carried out in this section to compare between four cases; no coding or interleaving, convolutional coding with code rate = 1/2 and K = 3, convolutional coding with block interleaving on a packet-by-packet basis, and convolutional coding with chaotic interleaving on a packet-by-packet basis. The results of these experiments are shown in Fig. 7.12.

Some sample received images are shown in Fig. 7.13. This figure reveals that the proposed chaotic interleaver on a packet-by-packet basis achieves the highest PSNR values with a convolutional code having K = 3.

Fig. 7.6 Received Cameraman image over a correlated fading channel at SNR = 20 dB with (**a**) PSNR = 31.1 dB, (**b**) PSNR = 31.5 dB, (**c**) PSNR = 32.1 dB, and (**d**) PSNR = 33.2 dB

7.1.5.4 Results Discussion

This section gives a comparison between the different Scenarios for image transmission over the ZigBee network based on the Number of Lost Packet (NLP %) and the PSNR. Figure 7.14 shows the variation of the NLP % with the channel SNR for the Cameraman image transmission over a correlated fading channel for the different scenarios. This figure indicates the efficiency of the proposed chaotic interleaver. As shown in this figure, the chaotic interleaver on a packet-by-packet basis is effective at low SNR values for short constraint length convolutional encoders.

Fig. 7.7 Received Cameraman image over a correlated fading channel at SNR = 30 dB with (**a**) PSNR = 39.1 dB, (**b**) PSNR = 41 dB, (**c**) PSNR = 41.1 dB, and (**d**) PSNR = 43.1 dB

Figure 7.15 shows the variation of the PSNR with the channel SNR for the Cameraman image transmission over a correlated fading channel for the different scenarios. It is clear that the received images using the chaotic interleaver on a packet-by-packet basis have the highest PSNR values. The proposed technique with K = 3 performs better than that with convolutional coding only with K = 7. In fact, the complexity increases exponentially with K. This means that the complexity of the decoder with K = 7 equals 16 times that of the decoder with K = 3. So, with the proposed chaotic interleaving technique on a packet-by-packet basis, the performance is improved, while the complexity is reduced.

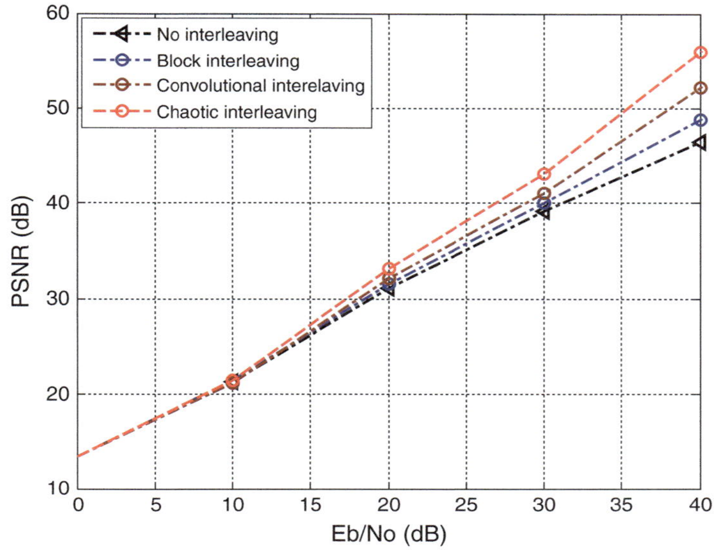

Fig. 7.8 PSNR vs. SNR for the received Camerman image over a correlated fading channel

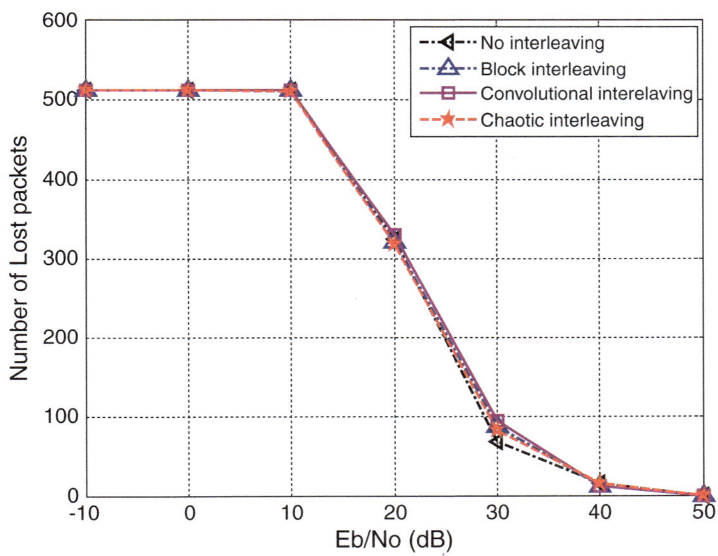

Fig. 7.9 Number of lost packets vs. SNR for the received Camerman image over a correlated fading channel

Fig. 7.10 Received Cameraman image over a correlated fading channel at SNR = 10 dB using the convolutional code with code rate = 1/2 and different constraint lengths, (**a**) PSNR = 23.12 dB, (**b**) PSNR = 23.35 dB, and (**c**) PSNR = 23.73 dB

7.1.6 The Scenario Recommendations

The scenario has presented a simple and efficient novel chaotic interleaver for the transmission of images over the ZigBee network. Also, it has studied the performance of the proposed interleaver with convolutional codes having different constraint lengths. A comparison study between the proposed interleaver and the conventional interleavers has been presented. The computer simulation results have revealed the effectiveness of the proposed interleaver at medium and high SNR

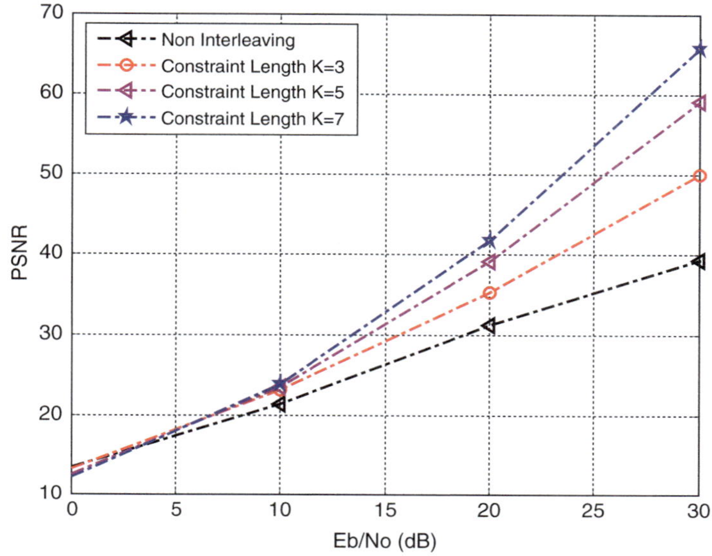

Fig. 7.11 PSNR vs. SNR for the received Camerman image over a correlated fading channel with different constraint lengths

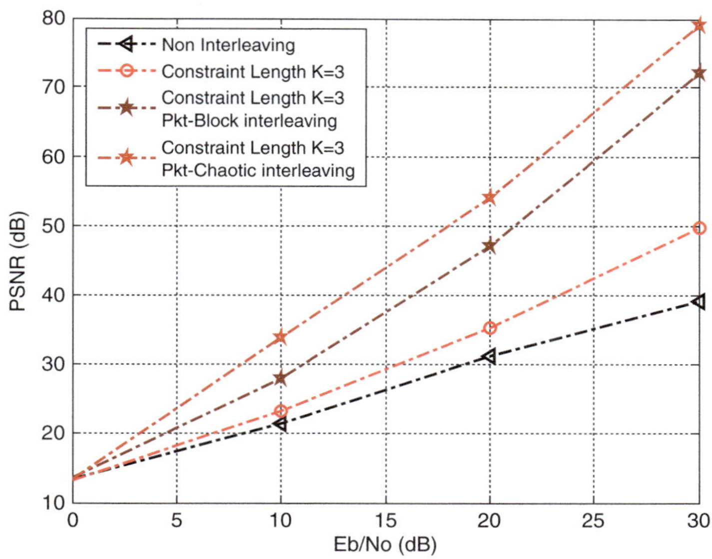

Fig. 7.12 PSNR vs. SNR for the received Camerman image over a correlated fading channel with different scenarios for image transmission

Fig. 7.13 Received Cameraman image over a correlated fading channel at SNR = 10 dB using the convolutional coding with code rate = 1/2 and K = 3, (**a**) PSNR = 23.12 dB, (**b**) PSNR = 27.8 dB, and (**c**) PSNR = 33.85 dB

values. The results have also proved that the chaotic interleaver improves the performance of the convolutional codes with short constraint lengths. It means the encoder with K = 3 with the chaotic interleaver performs better than the encoder with K = 7. In addition, the proposed scheme has the ability to enhance the security of the ZigBee network, because it permits the change of S_{key} from packet to packet. So, we can conclude that the proposed chaotic interleaving scheme on a packet-by-packet basis is suitable for ZigBee communication.

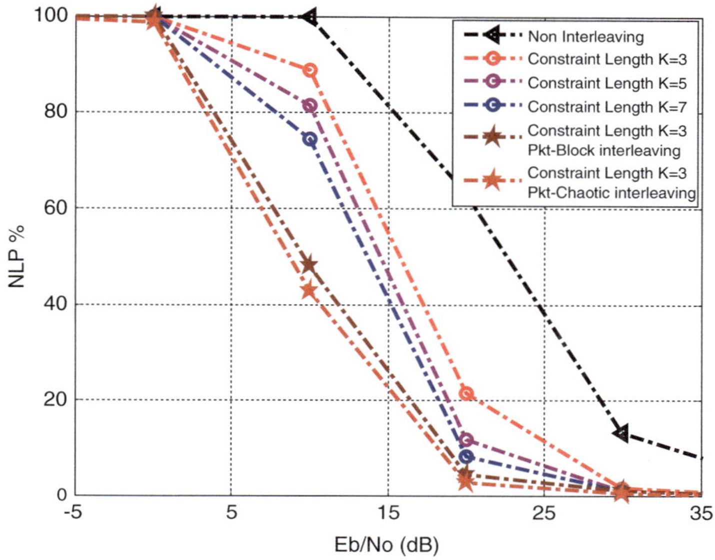

Fig. 7.14 NLP % vs. SNR for the received Camerman image over a correlated fading channel in the ZigBee network

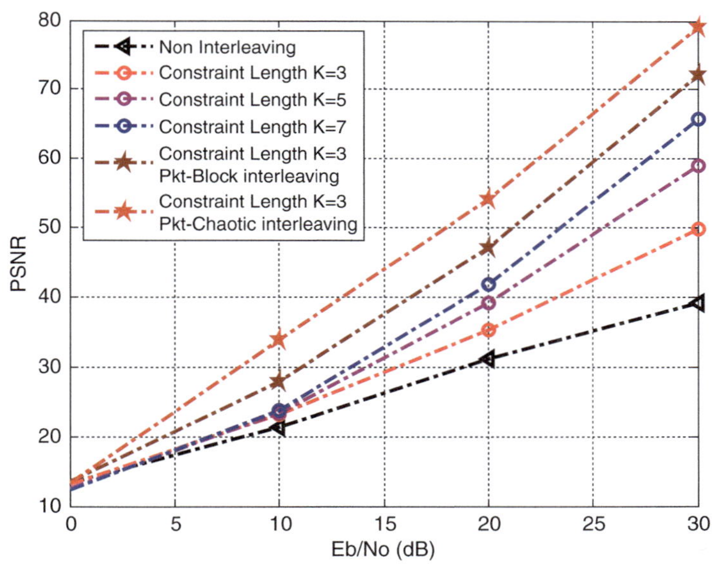

Fig. 7.15 PSNR vs. SNR for the received Camerman image over a correlated fading channel in the ZigBee network

7.2 Complexity Combating and Security Enhancing of WSN Networks

7.2.1 Simulation Scenario 2

The complexity of system Overshadow on the required consumed power and the time of processing. So, the reduction of complexity leads to the operations decreases, it is good to the WSN application and so good if meets the powerful performance.

7.2.1.1 Scenario Definition and Goal

This scenario aims to combat the complexity of longer constraint convolutional codes decoding though replacing it by powerful techniques. The expected that the proposed technique performs better and enhancing the security in addition. It also, studies the improving of colored JPEG image transmission over mobile Wireless Personal Area Network (WPAN) through the Bluetooth networks. The scenario uses many types of Enhanced Data Rate (EDR), Asynchronous Connectionless (ACL) packets. It presents a proposed chaotic interleaving technique for improving a transmission of colored images over burst error environment through merging it with error control scheme. The computational complexity of the used different error control schemes is considered. A comparison study between different scenarios of the image transmission is held in to choose an effective technique. The simulation experiments are carried over the correlated fading channel using the widely accepted Jakes' model. Our experiments reveal that the proposed chaotic interleaving technique enhances quality of the received colored image. In our simulation results show that the convolutional codes with longer constraint length are effective if its complexity is ignored. It reveals also that the standard error control scheme of old Bluetooth versions is ineffective in the case of colored image transmission over mobile Bluetooth network. Finally, the proposed scenarios of the standard error control scheme with the chaotic interleaver perform better than the convolutional codes with reducing the complexity.

7.2.2 Scenario Motivations

The WSN security is essential to success the applications. WSN may be implemented for national project touch the countries security. There is no probability for false alarm. The complexity of WSN should be low as much as possible to limit the power and limit the time. So, the power efficiency realized and latency decreased. These factors if met in one technique with satisfied performance, so the presented technique can be suitable and applicable within the WSN applications.

Wireless multimedia such as image transmission across the wireless networks has been gaining attention in the recent years because of the proliferation of technologies like Bluetooth, IEEE 802.11, 3G, and WiMAX [1]. This work presents many scenarios for image transmission over the mobile Wireless Personal Area Network (WPAN) through applying the Interleaved Forward Error Correction (IFEC) technique [2]. With increasing utilization of wireless devices, especially WPAN technologies applications, there are some important factors for all wireless systems power efficiency, efficient throughput, the Quality of Services (QoS) and the complexity.

The most of researchers were interested to apply the interleaved FEC technique on the convolutional codes [3, 4]. Also, the mobility probability of the WPANs nodes effects is not focused from the most researchers in this area. In this work, the interleaved FEC is presented with employing a powerful interleaver replaced the traditional interleavers [5]. The presented scenarios aimed to enhance the error performance of the block codes compared to the convolutional codes performance and reducing the need of complex error control codes. It is known that the complexity of the convolutional code and its decoder, Viterbi Algorithm (VA), are very high compared to the block codes such as Hamming codes. This scenario interested to improve the error control scheme of classic Bluetooth packets which is the Hamming code (15, 10) through using powerful data randomization tool as an interleaver [6, 7].

The interleaved FEC technique provides enhancing the error performance more than non-interleaved FEC. the scenario takes the complexity of the FEC schemes with the considerations, where it presents the results of the non-interleaved and interleaved Hamming code compared to the results of the convolutional codes with different constraint length (K). In general, the performance and the complexity of the error control schemes depends two main factors, the number of input of encoder data bit (k = data word length) and the number of output encoder bit (n = codeword length). In case of the convolutional codes, there is another important factor which is the constraint length of the convolutional encoder (K). The length of the K depends on the memory depth of the convolutional encoder (m), where $K = m + 1$. The effect of the K in the computational complexity of the VA is presented in Sect. III.

There are many fields merged the WPANs technologies in its applications. The most of the modern medical devices offer a kind of electronic data exchange interface. The wireless communications are available for medical services and health care equipments. The most common wireless technologies are very suitable for this purpose is ZigBee (IEEE 802.16.4) and Bluetooth (IEEE 802.16.1). In this scenario, we improve the efficiency of colored image transmission over Bluetooth network with proposing a mobile environment [8].

It operates within distance of 10–100 m. The structure of stations in different Bluetooth versions follows a piconet structure. Each piconet comprises up to seven Bluetooth devices working as slaves (S) and only one Bluetooth device as a master (M) station. A slave can be a member in more than one piconet. A master of any piconet may be a slave in other one. Up to 10 piconets can exist within Bluetooth range [9].

Bluetooth operates in the unlicensed 2.4 GHz ISM (Industrial Scientific Medical) frequency band, which is also utilized by various wireless and radio technologies, such as IEEE 802.11b/g standard [10], IEEE 802.16.4 standard, cordless telephones, and even microwave ovens. Bluetooth employs the Frequency Hopping Spread Spectrum (FHSS) technique to mitigate the interferences caused by other wireless services, coexisting in the 2.4 GHz frequency band. The Bluetooth technology presents the industrial specifications of wireless personal area networks (PANs), where it provides wireless media to connect and exchange information between devices. Bluetooth employs variable-size packets. These packets occupy different number of time-slots up to a maximum of five slots; each time-slot length is 625 μs. Bluetooth v. 2.1 has brought EDR packets types. These EDR packets support gross air rates of 2 Mbps and 3 Mbps through Л/4-DQPSK and 8DPSK modulation respectively [11, 12].

There are many of scenarios studied using Bluetooth in hospitals environments, control automations, and industrial application [13]. All previous proposed applications may need transmission colored images. In this scenario, we try to study the transmission of colored images over a mobile Bluetooth network and trying to improve the performance of colored image transmission. In our simulation, we use different techniques, such as interleaving techniques and error control technique. The transmission of image simulation is carried out using one of EDR Bluetooth packets [2DH3]. 2DH3 packet is one of ACL link packets. This packet carries 3,000 bits Payload (PL) uncoded data. In this scenario, we proposed using default 2DH3 packet and proposed packets through using error control codes schemes, such as convolutional code and Hamming code [14].

7.2.3 Scenario Preparations

The last case in our simulation is called chaotic interleaving. In this scheme, the column of digital image is used to be input of chaotic map. That is means the encryption process is carried over column by column, separately. After columns encryption the image is reconstructed and segmented to Bluetooth packets. The colored image of simulation is JPEG image 400 × 300 pixels, which is shown in Fig. 7.16. First step in our simulation is converting the image to forms (600 × 600) square matrix; a new image is shown in Fig. 7.17. Second step, converting the image to binary bits streams. The secret key of chaotic interleaving/encryption process, $S_{key} = (n_1, n_2, n_3, \ldots, n_m)$, for example the secret key can be as follows, $S_{key} = (10,5,12,5,10,8,14,11,\ldots, 10,5,12,5,10,8,14,11)$. The sum of the total S_{key} numbers must be equal 600 as shown in Fig. 7.18, it gives the encrypted image version within the dimensions (600x600). So, there are many secret keys can be

Fig. 7.16 Original natural view image

generated, unique key can be dedicated for every transmitted packet, that means powerful security addition. But in the truth, the form of secret key numbers effects on the degree of the bits of data/image pixels randomizing and what will the consequent also [28, 29].

7.2.3.1 Image Preparation and Fragmentation

Due to the communication channel effects the Bit Error rate (BER) is much higher and the burst error is expected [30]. Furthermore, the throughput may fluctuate due to time varying characteristics of wireless channel. This section studies the reasons of fragmentation and the number of segments in every case.

Our simulation experiments are carried out over the 2.4 GHz band in the case of Bluetooth with different cases simulations. Due to the limits of the packets size the fragment of image is important for complete the transmission. The size of the H. jpeg image 360×10^3 bytes, then the number of segment calculated from this equation

$$No.Segments\_1 = \frac{360 \times 10^3 \times 8(\text{Im}ageSize)}{2880(2DH_3BluetoothPacket)} \tag{7.1}$$

1. In case of Bluetooth system, length of packet is 256 bits in the case of uncoded PL of 2DH3 EDR packets [30].

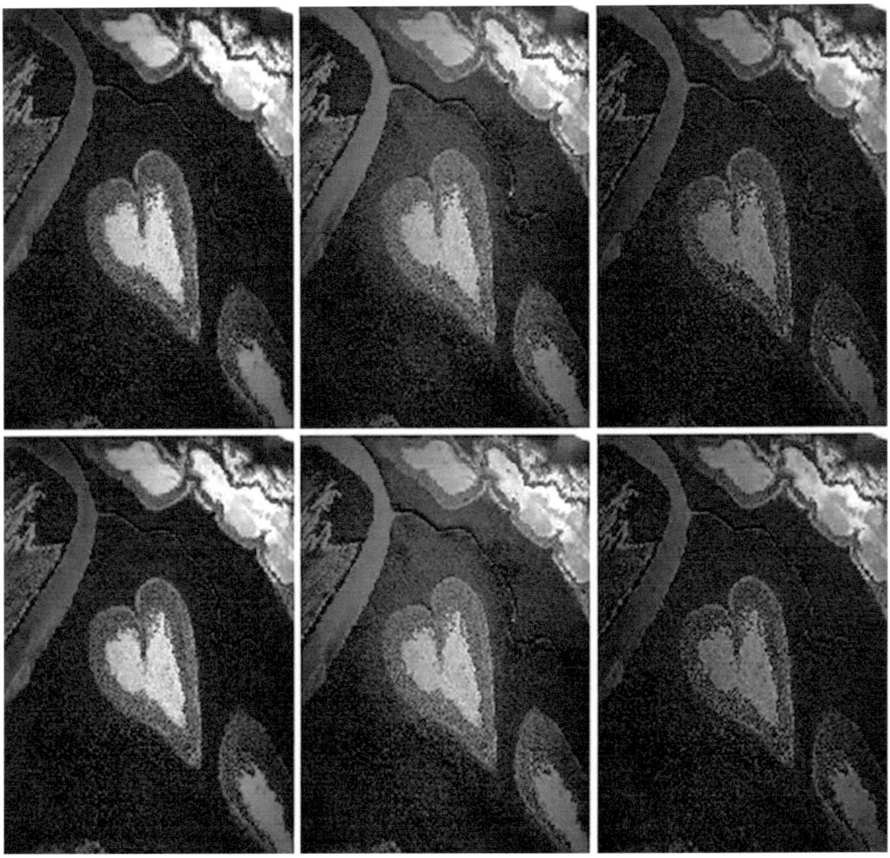

Fig. 7.17 The converted version of image

Fig. 7.18 The encrypted version of the image

$$No.Segments\_2 = \frac{360 \times 10^3 \times 8(ImageSize)}{1500(2DM_3BluetoothPacket)} \tag{7.2}$$

2. The case of encoded 2DH3 EDR which is called now 2DM3 Bluetooth packets the PL length of packet is half of previous case [Convolutional code] [19].

$$No.Segments\_3 = \frac{360 \times 10^3 \times 8(ImageSize)}{2000(2DM_3BluetoothPacket)} \tag{7.3}$$

The final case is 2DM3 packets with using Hamming code (15, 10). This error control scheme is a standard protection technique of classic Bluetooth packets [7, 8].

The length packets in our simulation cases are proposed for simplify the simulation and fragment the image.

7.2.3.2 Simulation Conditions

In this section, the simulation environment used for carrying out our experiments is described. An important assumption used in the simulation is that a packet is discarded if there is an error Bluetooth packet is discarded if there is an error in the header portion or the data field. This is a realistic assumption to simulate the real Bluetooth system operation [31].

In our simulation experiments, a mobile environment of Bluetooth terminal is taken in our consideration. That means, in our experiments Jakes model used in the case of correlated Rayleigh fading channel. The mobile Bluetooth device velocity is 10 mile/h, and the carrier frequency is 2.46 GHz. The Doppler spread is 36.6 Hz. The motion of the mobile terminal relative to fixed one and surrounding environment leads to a frequency shift between the transmitted frequency and the received one. The Doppler shift experienced by the mobile terminal is given by Eq. 7.11 [32]:

$$f_d = f_r - f_t = \frac{v}{\lambda}\cos\theta \tag{7.4}$$

f_t is the frequency of the transmitted signal.
f_r is the frequency of the received signal.
v is the speed of the mobile terminal or may be the relative velocity between two mobile terminals [33].
λ is the wavelength of the transmitted signal.

The maximum Doppler shift (f_m) is given at $\cos\theta = 1$, as shown in Eq. 7.12 [34].

$$f_m = \frac{v}{\lambda} \tag{7.5}$$

We will concentrate in our experiments on 2DH3 Bluetooth packets with length (standard length 3,000 bits) for transmitting the colored image, and the results can be generalized for other Bluetooth packets and other short range wireless systems. In our simulation the length of data field is changed according to the simulation case. The channel of simulation is correlated Rayleigh fading channel only for simulate the mobile Bluetooth terminal. The peak signal to noise ratio (PSNR) and number of packet loss (NPL) are evaluated between the original colored image and the received colored image [26].

The higher the PSNR and lower NPL mean the more efficient the Bluetooth system performance. In our work, MATLAB was used for carrying out our simulation experiments of different cases. The simulation results have been gotten by transmission of the image over different SNR values.

7.2.4 *Scenario Experiments*

In this section, the computer simulation results are presented. A correlated Rayleigh fading channel is used. The channel model utilized is the Jakes model [27, 35]. The assumed mobile Bluetooth device velocity is 10 miles/h, and the carrier frequency is 2.46 GHz. This carrier frequency is one of 79 Bluetooth carriers. The Doppler spread is 36.6 Hz. The original Cameraman image used in the experiments is a natural view image; it is given in Section 7.2.3, Scenario Preparations, pre-transmission process over the simulation program using Matlab. The computer simulation section contains three categories of experimental simulations.

In this section, different scenarios are employed for the image transmission over the Bluetooth network. The image binary sequence to be transmitted is fragmented into packets. The PSNR of the received images is used as an evaluation metric in this scenario.

7.2.4.1 Experiments Group 1

In the first experiment, the natural view image is transmitted using the uncoded $2DH_3$ packets over a correlated-fading channel with Signal-to-Noise Ratio $(SNR) = 5$ dB. Different scenarios of no interleaving, block interleaving, convolutional interleaving and chaotic interleaving are considered for comparison. The results of this experiment are shown in Fig. 7.19. From these results, it is clear that the effects of all interleaving schemes is approximately equal at low SNR values. Another experiment is repeated with $SNR = 15$ dB and the results are shown

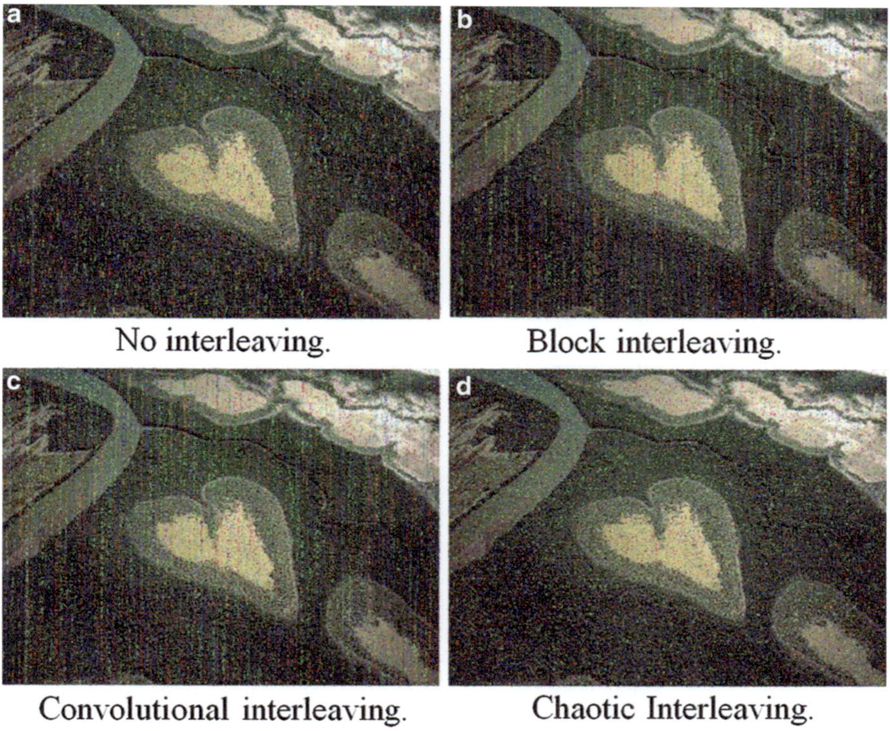

Fig. 7.19 Received image over a correlated fading channel at SNR = 5 dB with (**a**) PSNR = 16.7 dB, (**b**) PSNR = 16.8 dB, (**c**) PSNR = 16.7 dB, and (**d**) PSNR = 17 dB

in Fig. 7.20. From these results, we notice that the chaotic interleaver outperforms the other interleavers at moderate and high SNRs.

For the comparison purpose, the variation of the PSNR of the received image, correlation between the received and the original image (Cr) with the channel SNR are studied and the results are shown in Figs. 7.21 and 7.22. From these results, it is clear the chaotic interleaver enhancement begins at medium SNR values.

7.2.4.2 Experiments Group 2

In the following section of the computer simulation experiments, the non-interleaved and the interleaved 2/3 FEC using the Hamming code (15, 10) scheme is employed using the traditional block interleaver and the chaotic interleaver. The results of these experiments are given in Figs. 7.23, 7.24, and 7.25.

Figure 7.23 gives the some sample from the received image as shown in this figure, the non-interleaved FEC with non-powerful FEC scheme is inefficient

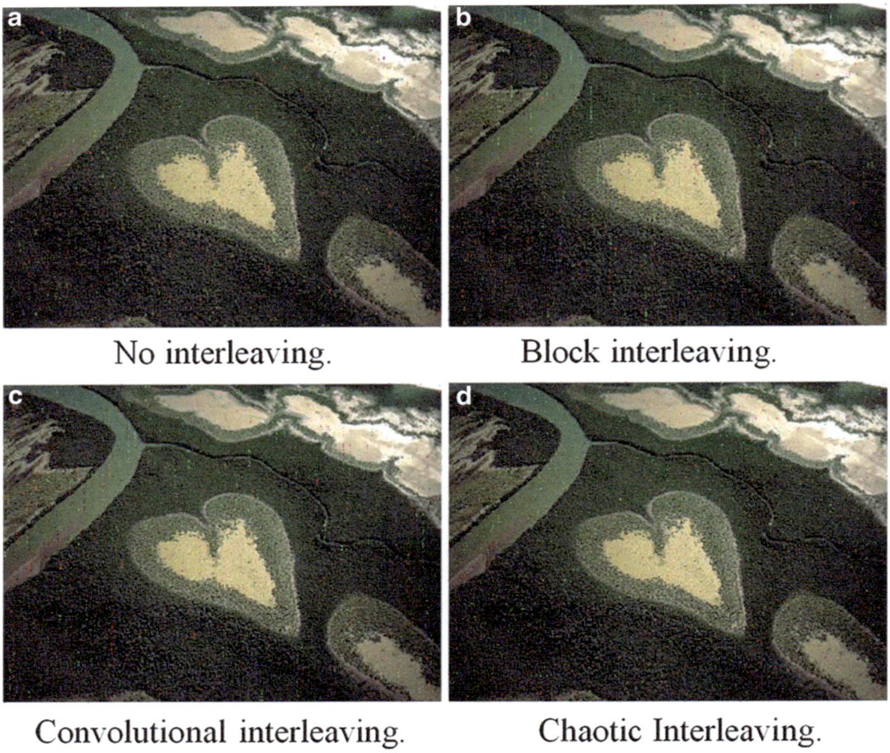

Fig. 7.20 Received image over a correlated fading channel at SNR = 15 dB with (**a**) PSNR = 26.3 dB, (**b**) PSNR = 26.4 dB, (**c**) PSNR = 26.4 dB, and (**d**) PSNR = 26.8 dB

technique for image transmission on mobile channel communications. Also, the interleaved Hamming code (15, 10) performs better than non-interleaved one. The proposed technique in this experiment, chaotic interleaved Hamming code (15, 10) enhances the PSNR of the received images more than the traditional one.

For the different scenarios comparison purpose, the variation of the PSNR of the received image, correlation between the received and the original image (cr) with the channel SNR are studied and the results are shown in Figs. 7.24 and 7.25. From these results, it is clear the interleaved block codes with chaotic interleaver leads to raise and refresh the weak block code capabilities.

In the previous sections, we mentioned that the complexity of the convolutional codes is much higher than the block codes. Also, it is mentioned that the main factor of the convolutional codes complexity is the constraint length (K). with the enhancement of the block codes (such as our case study Hamming code (15, 10)) through powerful interleaver, we may be do not need the complex error control schemes such as the convolutional codes with longer constraint length (K = 7 or K = 9) and reducing the complexity of the communications system with keeping the quality of transmission process.

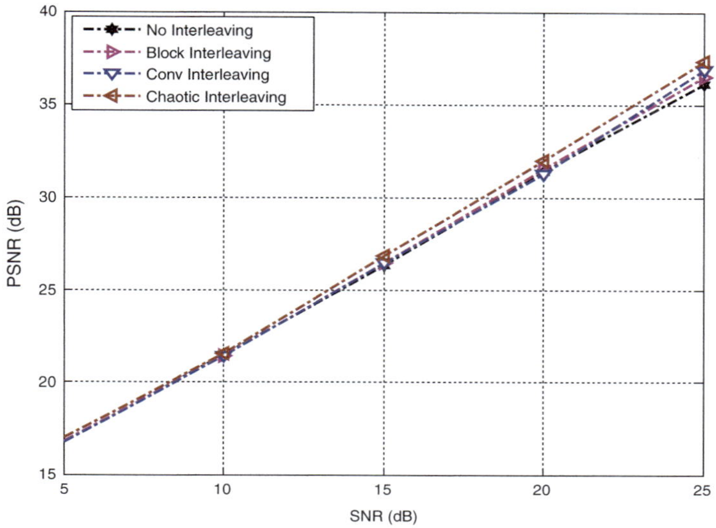

Fig. 7.21 PSNR vs. SNR for the image transmission over the correlated fading channel with different scenarios for uncoded packets

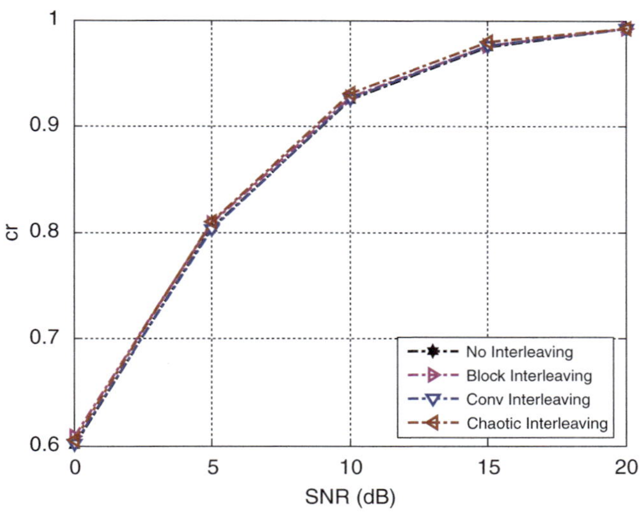

Fig. 7.22 Cr vs. SNR for the image transmission over the correlated fading channel with different scenarios for uncoded packets

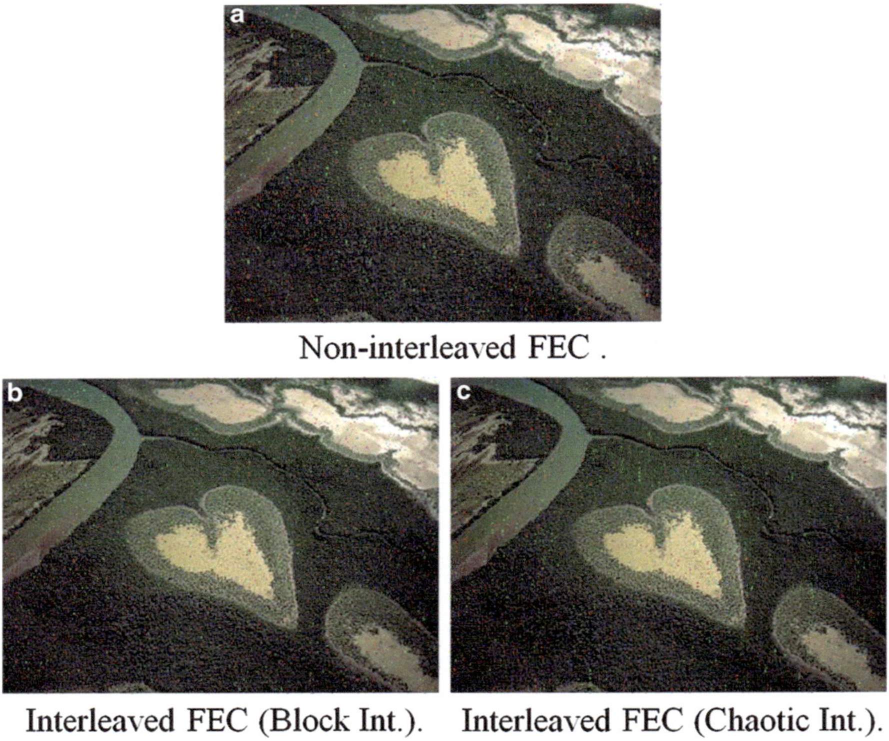

Fig. 7.23 Received image over a correlated fading channel at SNR = 10 dB with the non-interleaved and interleaved Hamming code (15, 10), (**a**) PSNR = 21.5 dB, (**b**) PSNR = 23.5 dB, and (**c**) PSNR = 26.6 dB

7.2.4.3 Experiments Group 3

To proof this expected results, in the following experiments we discuss the effect of the memory depth (m) of the convolutional codes on the performance of the image transmission on the mobile network. The constraint length depends on the memory depth (K = m + 1).

This experiment is devoted to investigate the effect of the constraint length K of the convolutional code on the image transmission process. The image is transmitted over a correlated fading channel having SNR = 10 dB using convolutional coding only with code rate = 1/2, and without interleaving. Constraint lengths of K = 3, 7, and 9 are investigated. Simulation results for this section are shown in Fig. 7.26. From these results, it is clear that the effect of the constraint length on the received image is marginal at SNR = 10 dB.

As shown in these figures the PSNR of the received image enhancement with the longer constraint length is not as expected. Also, it is clear from Fig. 7.27, that the large constraint length of the Convolutional encoder gives better PSNR

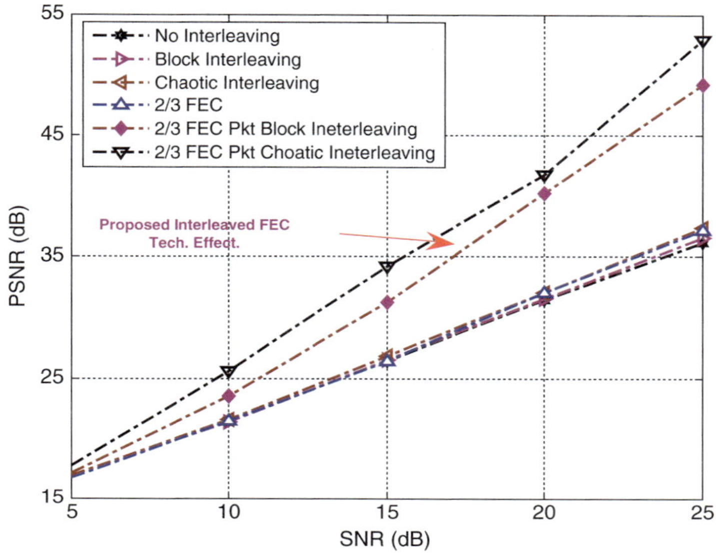

Fig. 7.24 PSNR vs. SNR for the image transmission over the correlated fading channel with different scenarios of interleaved FEC technique

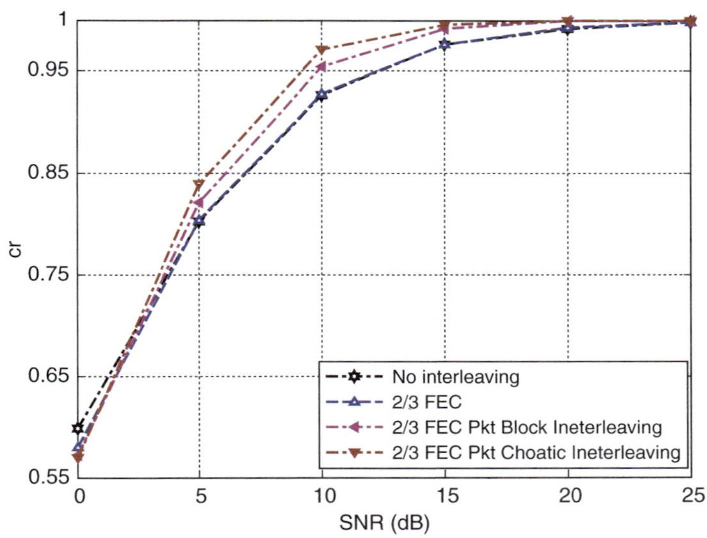

Fig. 7.25 Cr vs. SNR for the image transmission over the correlated fading channel with different scenarios of interleaved FEC technique

Code rate =1/2, K=3.

Code rate =1/2, K=7.

Code rate =1/2, K=9.

Fig. 7.26 Received image over a co related fading channel at SNR = 10 dB using the convolutional code with code rate = 1/2 and different constraint lengths, (**a**) PSNR = 23.5 dB, (**b**) PSNR = 23.2 dB, and (**c**) PSNR = 23.8 dB

performance, especially at medium and high SNR values in the channel at the expense of an increased complexity.

Figure 7.28 gives the correlation between the received image and the original one variation with channel SNR.

7.2.4.4 Results Discussion

In the following figures, the performance comparison of the proposed chaotic interleaved Hamming code (15, 10) and the convolutional code with the R = 1/2 and K = 3, 7, 9. Figure 7.29 gives the PSNR of the received image with the following scenarios of the image transmission, non interleaving uncoded packets, non-interleaved convolutional code with different K, and interleaved Hamming code (15, 10) with its traditional and the proposed constructions.

As shown in that figure, the proposed chaotic interleaved FEC with employing no-powerful error control scheme such as hamming code (15, 10) performs better than the longer constraint length convolutional codes.

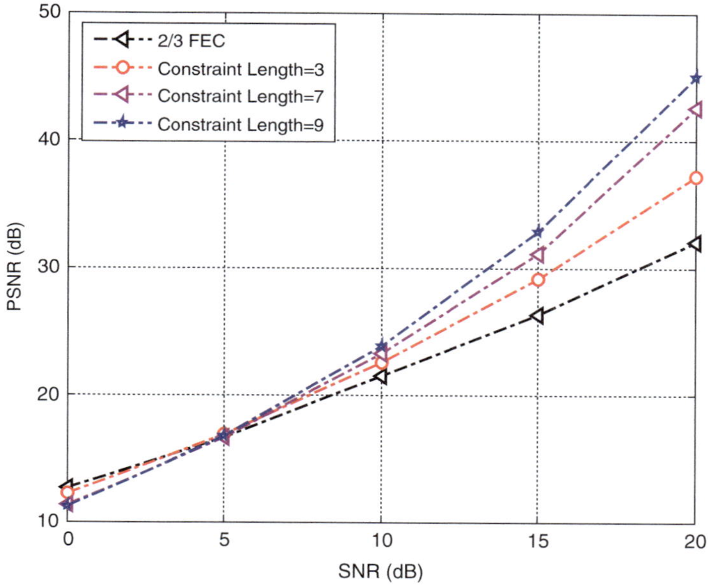

Fig. 7.27 PSNR vs. SNR for the received image over a correlated fading channel with different constraint lengths

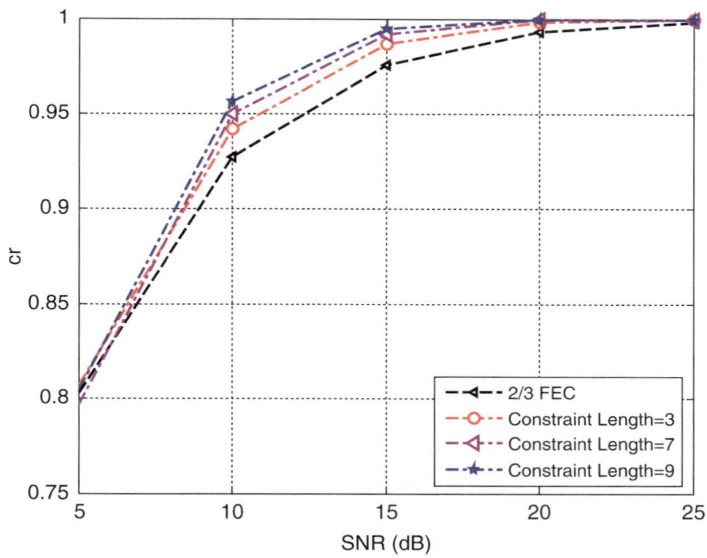

Fig. 7.28 Cr vs. SNR for the received image over a correlated fading channel with different constraint lengths

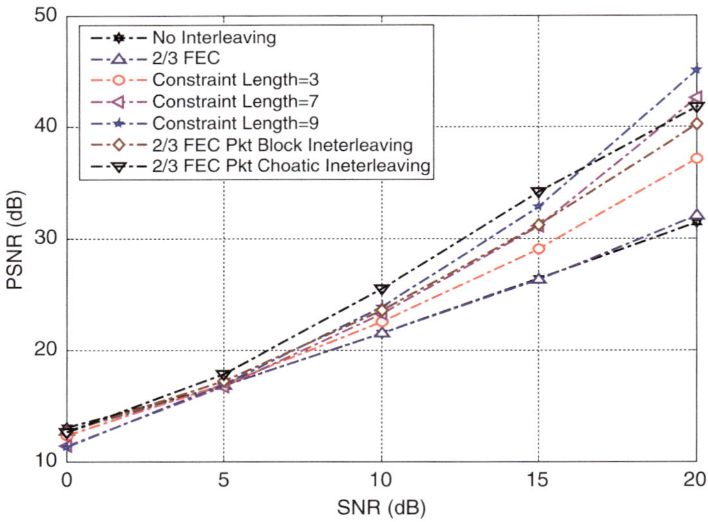

Fig. 7.29 PSNR vs. SNR for the received image over a correlated fading channel with different scenarios

Figure 7.30 gives the experiment results of the transmitted image over a correlated fading channel having SNR = 15 dB with employing the different previous scenarios.

As shown in the figure the highest PSNR is given with using the proposed chaotic interleaving technique with Hamming code (15, 10) to encode the 2DM3 packets. The previous results reveal that the need of high computational complexity error control scheme to enhance the image transmission over the mobile network can be restricted through merging powerful interleaver such as the chaotic interleaver with any weak block code.

The correlation between the received image and the original one of the previous experiment variation with the channel SNR is given in Fig. 7.31 for the different scenarios.

As shown in Fig. 7.32, the Number of Lost Packets (NLP) decreased with the encoded packets transmission. Also, the proposed chaotic interleaved FEC reduces the NLP packet and performs better than other scenarios.

Figure 7.33 gives the variation of the amount of data successfully transmitted (throughput) over the channel of simulation with channel SNR for different scenarios. As shown the proposed technique provides highest throughput over the channel.

The last metric which is used to evaluate the different scenarios performance is the error performance measurement through the Bit Error Rate (BER). Figure 7.34 gives the BER variation with different scenarios.

As shown in the previous Figs. 7.29, 7.30, 7.31, 7.32, 7.33, and 7.34, the using of chaotic interleaving technique as a tool for improving the colored image

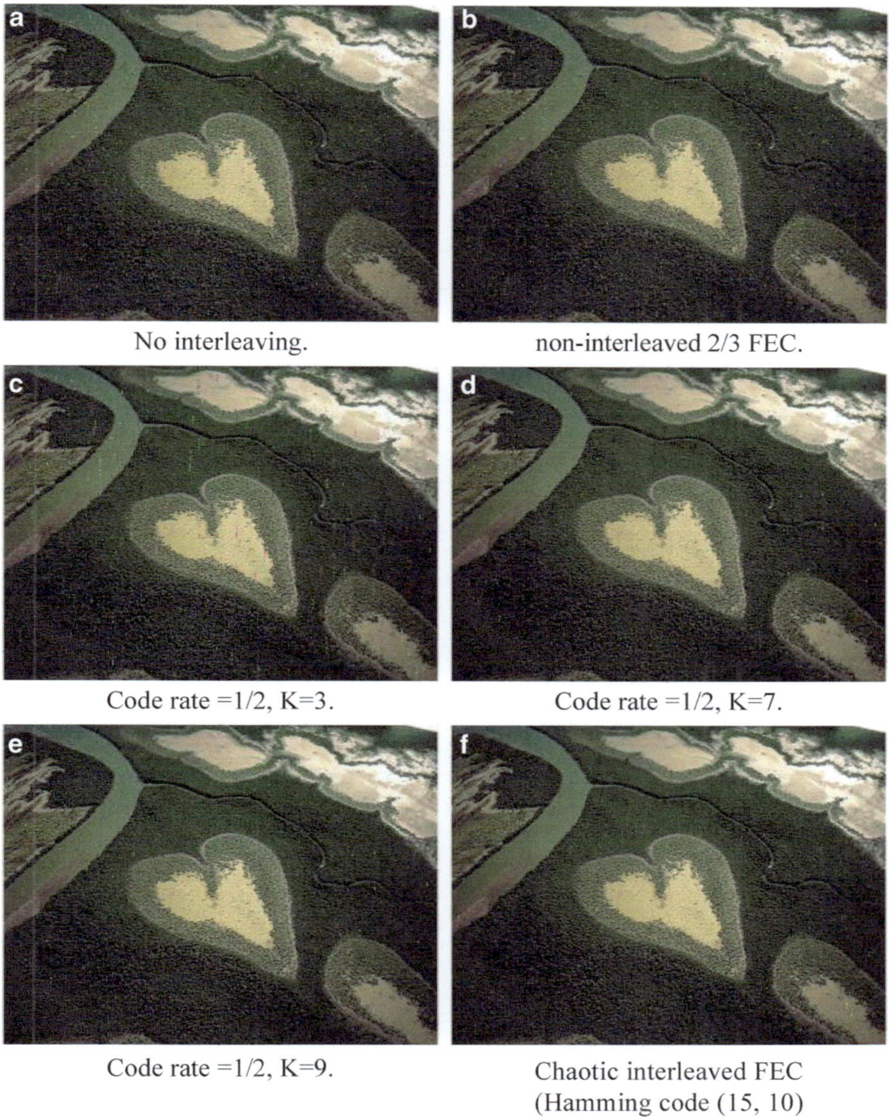

Fig. 7.30 The received colored image at SNR = 15 dB over the correlated fading channel with different scenarios: (**a**) PSNR = 26.3 dB, (**b**) PSNR = 26.35 dB, (**c**) PSNR = 29.1 dB, (**d**) PSNR = 31.0 dB, (**e**) PSNR = 32.8 dB, and (**f**) PSNR = 34.2 dB

transmission is ineffective manner, also at low SNR chaotic interleaving gives a good effect on received colored image. Standard error control scheme of classic Bluetooth packets gives a little improvement on received image. In the case convolutional code at K = 7 gives good results and decrease the NPL.

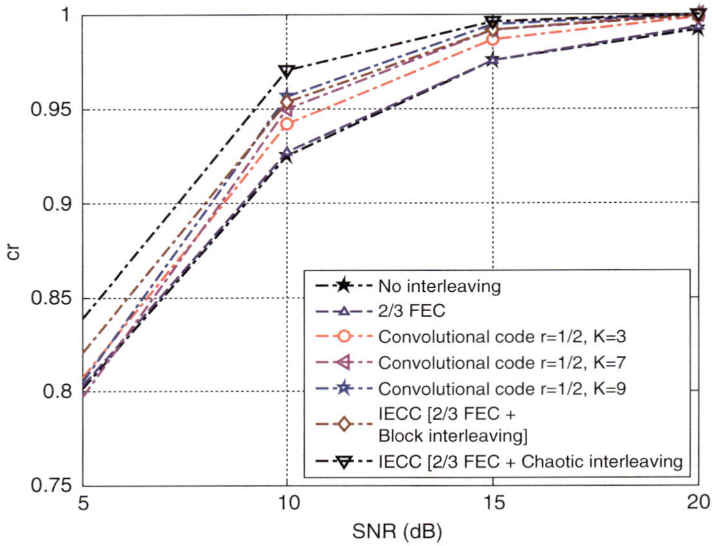

Fig. 7.31 Cr vs. SNR for the received image over a the correlated fading channel with different scenarios

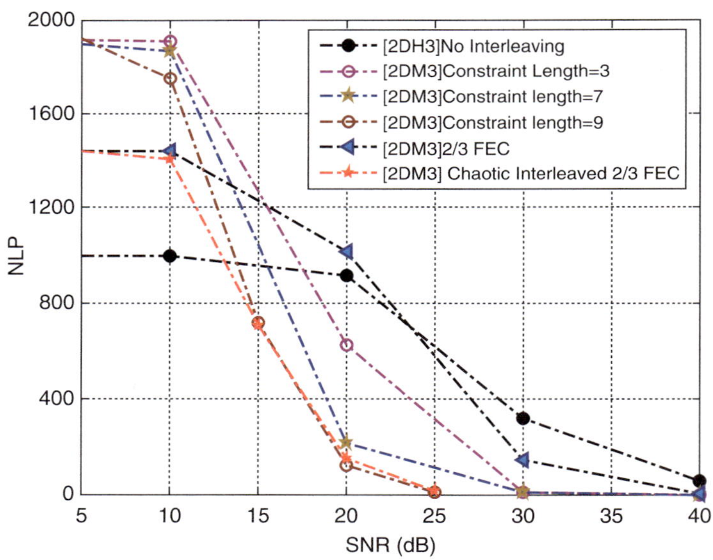

Fig. 7.32 NPL vs. SNR over the correlated fading channel using different scenarios for image transmission

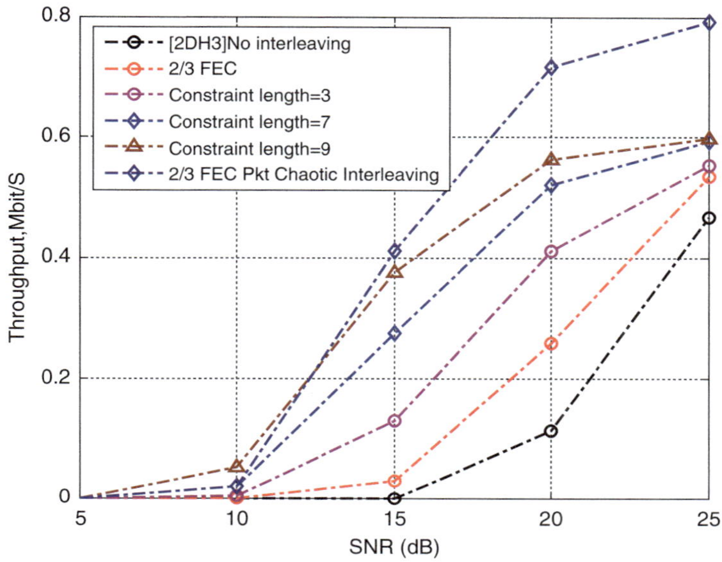

Fig. 7.33 Throughput vs. SNR over the correlated fading channel using different scenarios for image transmission

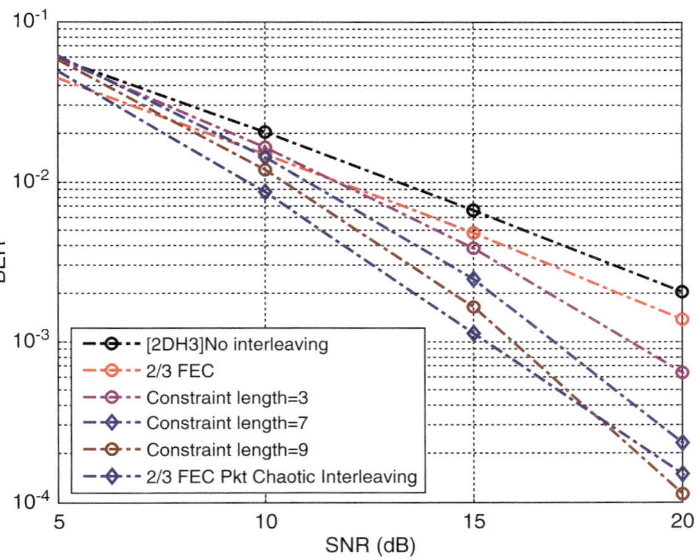

Fig. 7.34 BER vs. SNR over the correlated fading channel using different scenarios for image transmission

7.2.5 Scenario Recommendations

The scenario has presented a simple and efficient novel chaotic interleaver for the transmission of images over the WPAN Bluetooth network. Also, it has studied the performance of the proposed interleaver with standard error control code of old Bluetooth versions compared to the convolutional codes having different constraint lengths. A comparison study between the proposed interleaver and the conventional interleavers has been presented. The computer simulation results have revealed the effectiveness of the proposed interleaver at medium and high SNR values. The results have also proved that the chaotic interleaver improves the performance of the Hamming code (15, 10). It means the weak block code with the chaotic interleaver performs better than the convolutional codes with $K = 3$, 7, and 9. In addition, the proposed scheme has the ability to enhance the security of the Bluetooth network, because it permits the change of S_{key} from packet to packet. So, we can conclude that the proposed chaotic interleaving scheme on a packet-by-packet basis is suitable for mobile personal communication. Also, it gives a high quality image and eliminates the need for a complex of convolutional encoder.

7.3 Secured Medical Wireless Services Employing Watermarking Technique

7.3.1 Simulation Scenario 3

The field applications of different wireless technologies became the most essential one. The Tremendous development of the different wireless technologies is employed for the purposes of services and advanced medical care such as the Wireless Body Area Network (WBAN) IEEE 802.16.6.

7.3.1.1 Definition and Goal

In this research work, the coded secured wireless medical service technique is presented. The proposed scenario based on the audio watermarking technique is employed to embedding the medical inspection images, which are encrypted or unencrypted according to its level security needed. The audio watermarking in this research used the Singular Value Decomposition (SVD) mathematical technique to embed the watermark image into the audio signal. The used audio signal represents the medial inspection report belong these embedded medical images. In general, the watermarking technique is the art of embedding a piece of information into a cover signal. In the presented scenario, the piece information is a medial image and the cover signal is the medical report. The proposed approach can be used for providing multi-level security wireless link through using turbo coding with different rates by using puncturing techniques and different data frame lengths.

7.3.2 Scenario Motivation

With the rapid advancements of wireless communication and semiconductor technologies, the area of sensor network has grown significantly supporting a range of applications including medical and healthcare systems. A Wireless Body Area Network (WBAN) is a special purpose sensor network designed to operate autonomously to connect various medical sensors and appliances, located inside and outside of a human body. A WBAN system can offer two significant advantages compared to current electronic patient monitoring systems. The first advantage is the mobility of patients due to use of portable monitoring devices [1]. Second advantage is the location independent monitoring facility.

A WBAN node being an autonomous device can search and find a suitable communication network to transmit data to a remote database server for storage. It is also possible that a WBAN will connect itself to the internet to transmit data in a non-invasive manner. The health care sector is increasingly looking for advanced ICT (Information and Communication Technology) systems to efficiently administer the healthcare delivery for a range of services. Advanced ICT systems will be able to deliver healthcare not only to patients in hospitals and medical centers; but also in their homes and workplaces thus offering cost savings, and improving the quality of life of patients [2, 3].

Over bad communication channels, different Forward Error Correction (FEC) schemes are used. In this research, punctured turbo coding is used as a FEC technique with different code rates to perform UEP for different channel conditions and different frame lengths. It can significantly improve the performance of communication systems [4]. Our study focus to introduce a powerful and secure system for patient data. Watermarking technique is used to embed medical images for the patient into the audio file report. In fact, the environment of mobile communications suffers from long burst errors, our project presents a proposed technique depends on data randomizing tool to combat the burst errors. The proposed work aims at achieving powerful and secured medical service area through the presented scenarios utilizing the proposed technique. The work presents the proposed technique for Embedding important images in medical audio report using watermarking technique for security purposes. Also, the FEC can improve the quality of extracted the medical images, this point is considered in the proposed scenarios. For this purpose, the computer simulation experiments employed the widely used convolutional codes and the simple block code Hamming code. The simulation results have revealed that the proposed technique is efficient for wireless medical services.

The rest of scenario is organized as follows. The brief of WBAN concepts is given in section 7.3.3. In section 7.3.4, the WBAN and WSN relation is discussed.

The presented simulation scenarios are described in section 7.3.5. Finally, the simulation experiments are presented in section 7.3.8.

7.3.3 WBAN

WBANs networks are concerned with the remote monitoring of key human body physiological signals. It is oriented wireless networks, it is main purpose the medical services and human body care applications. So, its systems need to maintain high levels of reliability and low levels of latency as well as the high level security. These networks are based on the sensing process as the Wireless Sensor Network (WSN). The sensor devices have relatively low data rates but energy efficiency and node battery longevity are fundamental to enable devices to maintain connectivity for long periods of time. Patients are likely to be mobile and WBANs will be processing key sensitive data that must be securely processed. Nodes are likely to be worn continuously. Patients must have confidence in the technology and be comfortable with the device on a daily basis. The WBANs applications suffer from many constraints such as resource constraints, which include the energy, bandwidth, memory, limited transmission power.

WBANs must incorporate adequate levels of security and privacy to protect the sensitive nature of personal and medical related data. WBANs are vulnerable to the same forms of attacks as other wireless networks such as eavesdropping and Denial of Service (DoS). These threats can expose a WBAN to serious security vulnerabilities and affect its performance at the physical, data link, network, and transport layers [18]. Patient data must be stored locally and securely transmitted to the server side in the context of transmitting data to a remote health professional. WBAN designers need to ensure that the processes for system setup and day to day operation do not introduce security risk as these tasks will be undertaken by non technical personnel. Efficient cryptographic techniques need to be used to mitigate the risk of security breaches but must not add significant overhead to packet size or transmission times.

The WBAN applications targeted by the IEEE 802.16.6 standard are divided into medical and non-medical applications as given in Fig. 7.35. Medical applications include collecting vital information of a patient continuously and forward it to a remote monitoring station for further analysis.

7.3.4 WBAN and WSN Relation

As shown in Fig. 7.36, the section of environment terms as a specific area similar the grading of wireless technologies coverage area. The WSN applications may cover all these technologies. The WBAN applications concern the first and the second area. The first area which is the WBAN concerns the medical applications.

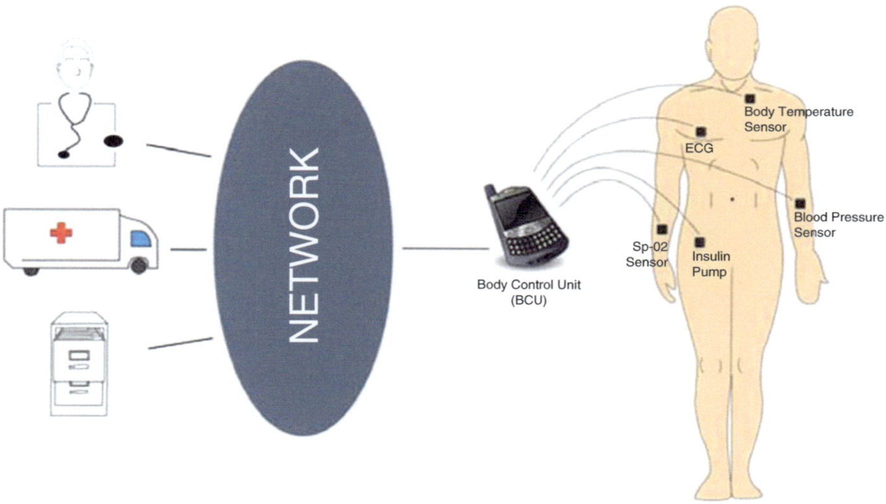

Fig. 7.35 Wireless medical and non medical applications

The other applications include another wireless technologies concerns the non-medical applications. The main shared line between the WSN and WBAN network is the basis of establishing of them which is the sensing function [35].

7.3.5 The Proposed Scenario Description

In this section, the proposed technique is described with the different scenarios of the simulation are used in the medical inspection image. This work aims to present secured wireless link of the medical service through the audio watermarking technique using the mathematical SVD technique. The embedded inspection medical images can be embed using the encryption or without according to the need. Also, the transmission process of the watermarked signal packets carried out using different scenarios. The scenarios include the different error control schemes and the long of transmitted packets effects. Figure 7.37 shows the stages and multi-level secured technique over the WBAN wireless link and medical image inspection processing stages.

The description contains the steps of medical images preparation and segmentation also, the encoding role description, audio report preparation, embedding process, transmission, channel, packet decoding, packet recollecting, medical image extracting, and evaluation.

The proposed scenarios include different techniques, multi-FEC options, different packet lengths, and normal and encrypted embedded images with the multi-secret key capability. The mentioned four techniques are proposed for investigate

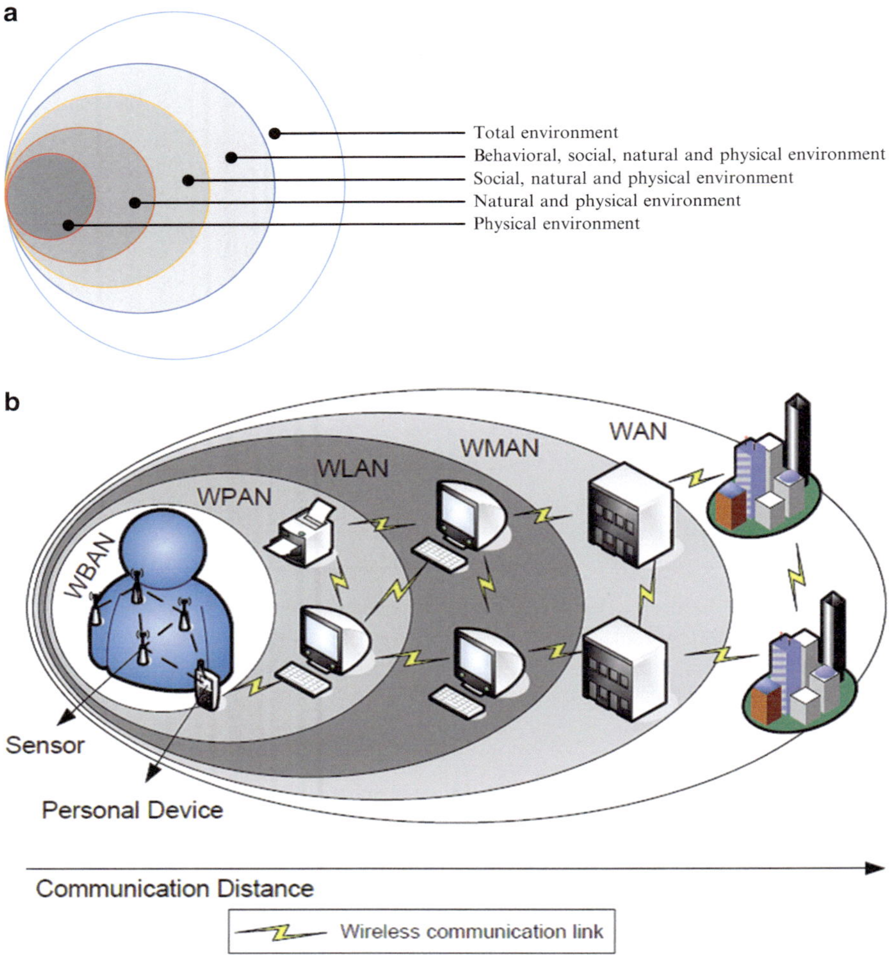

Fig. 7.36 The environment and wireless technologies similarity. (**a**) The environment sections structure. (**b**) The different wireless networks technologies

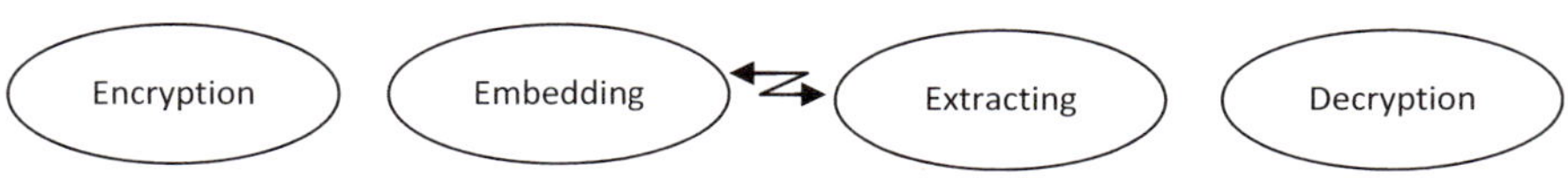

Fig. 7.37 The simulation steps and medical image transmission stages

two objects, which are the following: efficient quality of extracted medical images and multiple security levels based on the transmitted data needs. The computer simulation experiments cover the following points in brief:

- Multi-secret key for the encrypted embedded medical images.
- The testing of the proposed scenarios by the Matlab standard woman image.
- X-Ray inspection image of human head experiments.
- MRI inspection image of human brain experiments.
- Unequal packet length transmission.
- The FEC merging to enhance the extracted embedded medical images

Based on the two dimension chaotic encryption Baker map, in the following the different secret keys are employed within the presented scenario.

The list of secret key (S_{key}) in the below as a part from the Matlab code of the scenario:

```
Select=input('Press 1 if you need encryption');
if Select==1
 K_1_%n=[10,5,12,5,10,8,14,10,5,12,5,10,8,14,10,5,12,5,10,8,14,
10,5,12,5,10,8,14]
 K_2_ %n=[23,25,10,6,23,25,10,6,23,25,10,6,23,25,10,6]
 K_3_% n=[50,14,50,14,50,14,50,14]
K_4_ %n=[50,50,6,50,50,50]
K_5_ % n=[32,64,32,64,32,32]
K_6_ %n=[64,128,64]
 %%%%%%%%%%%%%%%%%%%$$$$$$$$$$$$
K_7_ %n=[256]
K_8_ %n=[128,128]
K_9_ %n=[64,64,64,64]
K_10_ %n=[32,32,32,32,32,32,32,32]
K_11_ %n=[16,16,16,16,16,16,16,16,16,16,16,16,16,16,16,16]
K_12_ %n=[8,8,8,8,8,8,8,8,8,8,8,8,8,8,8,8,8,8,8,8,8,8,8,8,8,8,8,
8,8,8,8,8]
K_13_%n=[4,4,4,4,4,4,4,4,4,4,4,4,4,4,4,4,4,4,4,4,
4,4,4,4,4,4,4,4,4,4,4,4,4,4,4,4,4,4,4,4,4,4,4,4,4,4,4,4,4,4,
4,4,4, 4,4,4,4,4,4,4,4,4,4,4]
K_14_%n=[2,2,2,2,2,2,2,2,2,2,2,2,2,2,2,2,2,2,2,2,2,2,2,2,2,2,2,
2,2,2,2,2,2,2,2,2,2,2,2,2,2,2,2,2,2,2,2,2,2,2,2,2,2,2,2,2,2,2,
2,2,2,2,2,2,2,2,2,2,2,2,2,2,2,2,2,2,2,2,2,2,2,2,2,2,2,2,2,2,2,
2,2,2,2,2,2,2,2,2,2,2,2,2,2,2,2,2,2,2,2,2,2,2,2,2,2,2,2,2,2,2,
2,2,2,2,2,2,2,2]
 K_15_% n=ones(1,256);
 K_16_ %n=[10,17,5,18,14,10,17,5,18,14,10,17,5,18,14,10,17,5,
18,14];
K_17_%n=[4,1,5,6,4,1,5,6,4,1,5,6,4,1,5,6,4,1,5,6,4,1,5,6,4,
1,5,6,4,1,5,6,4,1,5,6,4,1,5,6,4,1,5,6,4,1,5,6,4,1,5,6,4,
1,5,6,4,1,5,6,4,1,5,6]
   [pr,pc]=chaomat(n);
   W=chaoperm(W,pr,pc,3,'forward');
end;
```

The constraints consideration is essential point in the wireless sensing networks successful applications. Such as the memory constraint is related to the security level over these networks. The multi-security keys technique for example needs additional memory. The memory is not problem in the sink station (Base station), so it may play this role and manage the secret key changing between the sensing stations in its network.

7.3.6 The Proposed SVD Audio Watermarking Approach

The SVD algorithm is exploited in image watermarking [22–24]. In this scenario, this idea is extended into audio watermarking. The SVD mathematical technique provides an elegant way for extracting algebraic features from a 2-D matrix. The main properties of the matrix of the SVs can be exploited in audio watermarking. When a small perturbation happens to the original data matrix, no large variations occur in the matrix of singular values, which makes this technique robust against attacks [22–24].

7.3.6.1 SVD Technique

The SVD mathematical technique provides an elegant way for extracting algebraic features from a 2-D matrix as shown in Fig. 7.38 (by Beltrami in 1873 and Jordan in 1874). It extended to rectangular matrices by Eckart and Young in the 1930s.

U and V are orthogonal matrices, and S is a diagonal matrix consisting of singular values.

- SVD is one of the most useful tools of linear algebra with several applications to WSN.
  - Image compression
  - Watermarking
  - Other signal processing applications

At the heart of our technique is the general singular value decomposition theorem for matrices. Any $m \times n$ matrix A with real entries can be written as a linear combination

$$A = \sum_{k=1}^{r} \sigma_k u_k v_k^T \tag{7.6}$$

where r is the rank of A. The σ_k are positive numbers, ordered from greatest to smallest, called the *singular values* of A; the vectors u_k are orthonormal vectors in R^m (called *left singular vectors*); and the vectors v_k are orthonormal vectors in R^n (*right singular vectors*). Each *outer product* $u_k v_k^T$ is an $m \times n$ matrix of rank one

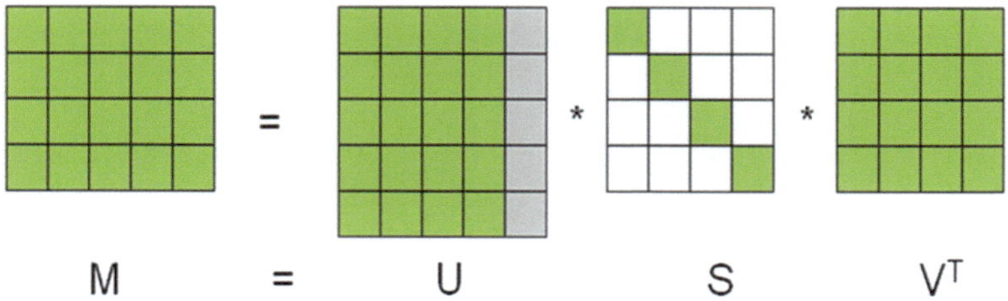

Fig. 7.38 The simulation steps and medical image transmission stages

formed by ordinary matrix multiplication of the column vector u_k by the row vector, v_k^T.

The SVD is a mathematical technique. It can be seen as a method for transforming correlated variables into a set of uncorrelated ones that better expose the various relationships among the original data items. At the same time, SVD is a method for identifying and ordering the dimensions along which data points exhibit the most variation. The SVD is linear algebra useful in several signals processing application.

SVD can be looked from the image processing perspective in [36], an image can be viewed as a matrix with nonnegative scalar entries. The SVD of an image A with size m × m is given by $A = USV^T$, where U and V are orthogonal matrices, and $S = \text{diag}(\lambda_i)$ is a diagonal matrix of singular values λ_i, $i = 1, 2, \cdots, m$, arranged in decreasing order. The columns of U are the left singular vectors, whereas the columns of V are the right singular vectors of the image A. This process is known as the Singular Value Decomposition (SVD) of A, and can be written as

$$A = USV^T \tag{7.7}$$

$$A = [u_1, u_2, \cdots u_m] \times \begin{pmatrix} \lambda_1 & 0 & \cdots & 0 \\ 0 & \lambda_2 & \cdots & 0 \\ \vdots & 0 & \ddots & 0 \\ 0 & 0 & \cdots & \lambda_m \end{pmatrix} \times [v_1, v_2, \cdots, v_m]^T \\ = \sum_{i=1}^{r} \lambda_i u_i v_i^T \tag{7.8}$$

where r is the rank of A, u_i and v_i are the left and right singular vectors, respectively. It is important to note that the singular values specify the luminance of the image, whereas the corresponding pair of singular vectors specifies the geometry of the image.

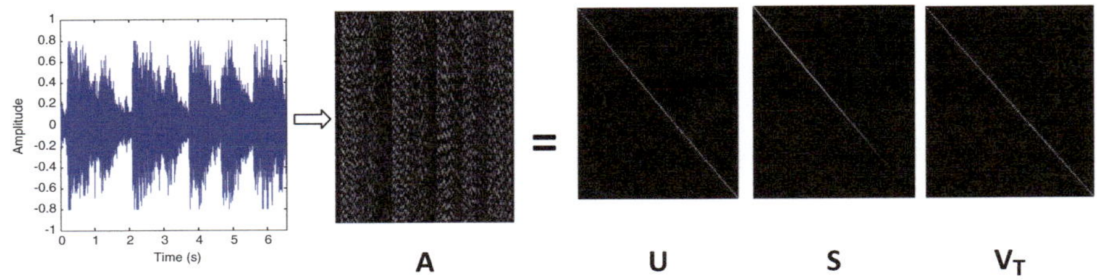

Fig. 7.39 The original audio signal and the SVD contents

7.3.6.2 Watermark Embedding

In this section the steps of the medical images embedding are described. As shown in Fig. 7.39, the original audio signal is shown, this audio signal represents the audio reports about the embedded medical images or any audio signal. The utilized audio signal may be song section for cover the important and high secured medical images or the embedded medical reports. The audio signal is converted to form square matrix within 256 × 256 dimensions, A. So, the SVD mathematical technique will carried out on this square matrix to produce three parts, U, S, and V^T.

Figure 7.39 gives the contents of the encrypted watermarking signal. The steps of the proposed SVD audio watermark embedding algorithm are summarized as follows:

1. The 1-D audio signal is transformed into a 2-D matrix (A matrix).
2. The SVD is performed on the A matrix.

$$A = USV^T \tag{7.9}$$

3. The chaotic encrypted watermark (W matrix) is added to the SVs of the original matrix.

$$D = S + KW \tag{7.10}$$

A small value of K of about 0.01 is required to keep the audio signal undistorted.

4. The SVD is performed on the new modified matrix (D matrix).

$$D = U_w S_w V_w^T \tag{7.11}$$

5. The watermarked signal in 2-D format (A_w matrix) is obtained using the modified matrix of SVs (S_w matrix).

$$A_w = US_w V^T \tag{7.12}$$

6. The 2-D A_w matrix is transformed again into a 1-D audio signal.

7.3.6.3 Watermark Detection

To extract the possibly corrupted watermark from the possibly distorted watermarked audio signal, given U_w, S, V_w matrices and the possibly distorted audio signal, the above steps are reversed as follows:

In the process of extracting, the side information Uw and Vw are error free. These contents are used in the algorithm as the same original. With the proposed the channel errors are free.

1. The 1-D audio signal is transformed into a 2-D matrix A_w^*. The * refers to the corruption due to attacks.
2. The SVD is performed on the possibly distorted watermarked image (A^*_w matrix).

$$A_w^* = U^* S_w^* V^{*T} \tag{7.13}$$

3. The matrix that includes the watermark is computed.

$$D^* = U_w S_w^* V_w^T \tag{7.14}$$

4. The possibly corrupted encrypted watermark is obtained.

$$W^* = (D^* - S)/K \tag{7.15}$$

5. The obtained matrix W^* is decrypted.
6. The correlation coefficient between the decrypted matrix and the original watermark is estimated. If this coefficient is higher than a certain threshold, the watermark is present.

Figure 7.40 shows the previous steps of the watermark extraction within applicable example of embedded X-ray medical inspection image extraction. The first step of the extraction is decryption of the encrypted watermark signal. So, the extraction of embedded image is starting and separating the audio signal. To ensure the extracted image quality, there are some metrics should be utilized such as the correlation between the original and the extracted medical image.

7.3.7 *Objective Quality Metrics*

There are some of metrics are devoted to evaluate the algorithm performance and suitability and its applicability on the medical images. First, the standard Matlab image "woman" is utilized to test the algorithm performance. The results of this test are obtained by carrying out the computer simulation experiments using different transmission of watermarked signal scenarios. Also, the Peak Signal to Noise Ratio

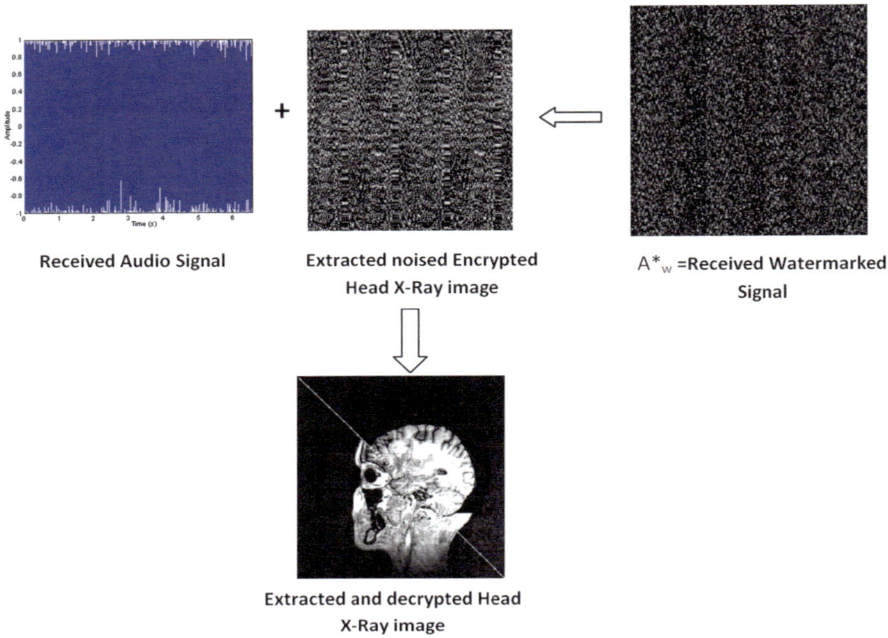

Fig. 7.40 The extracting process from the SVD contents, example of AWGN channel with SNR = 0 dB uncoded packets, Uw, Vw are error free contents

(PSNR) and the correlation between the original medical images and the extracted of the received images are used to evaluate the extracted images quality.

7.3.8 Simulation and Evaluation

In the section of the computer simulation experiment, there many group of separate experiments. The section of experiment manipulates many techniques aim to enhance the extracted medical images quality. The multi-codes technique is utilized to encode the transmitted packets of audio watermarked signal. The experiments also, tested the role of the codes. Also, the different lengths of the transmitted packets of the watermarked signal effects on the quality of extracted medical images are presented under the title of unequal packet lengths idea. The last one is the multi-level security. This idea is presented in two methods, first, the embedded medical image may encrypt or not according to the need. The second method, the experiments presented with multi-secret key with clearing the effects of these key on the encrypted and extracted medical images. These groups are different in the simulation scenarios. There several applications for the proposed scenarios of the audio watermarking techniques are presented. The first group of experiments is devoted to evaluate the presented multi-codes multi-level security technique based on the audio watermarking using the standard image before testing the proposed scenarios for the medical images.

7.3.8.1 Multi-key

In this section, several secret key S_{key} is presented. The Keys graded from the simpler to more complexes, the effects of these key changing are cleared on the encrypted medical images and the extracted ones.

There are three images are utilized in this work. The first one is the Matlab standard Woman image. Figure 7.41 gives the original woman image standard and the two encryption versions.

There are two medical images; one is the head X-Ray image. Figure 7.42 shows the original and the two encryption versions of it.

The last medical image is utilized in the computer simulation experiments for evaluating the proposed scenarios is the MRI image of the human brain. The original and the encrypted versions of this image are given in Fig. 7.43.

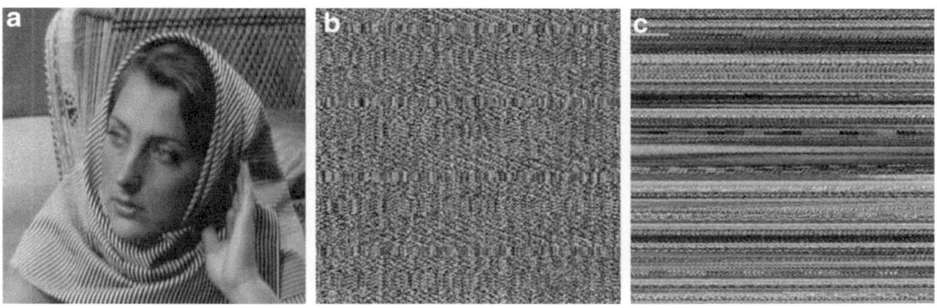

Fig. 7.41 Woman image, (**a**) original, (**b**) encrypted version with K1, (**c**) encrypted version with K11

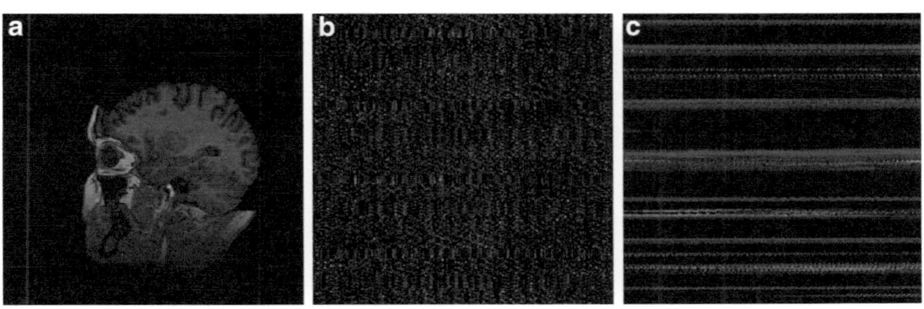

Fig. 7.42 X-Ray image, (**a**) original, (**b**) encrypted version with K1, (**c**) encrypted version with K11

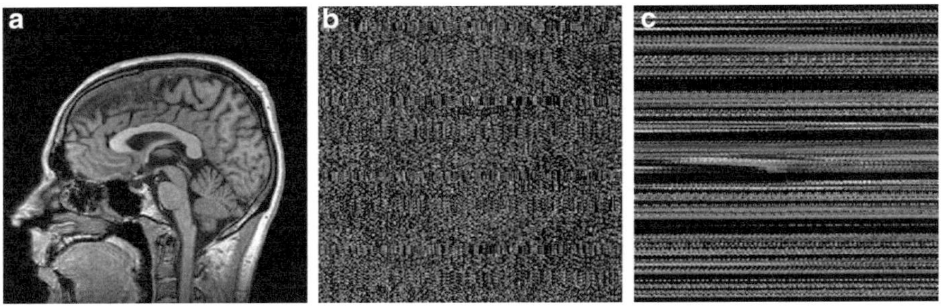

Fig. 7.43 MRI-Brain image, (**a**) original, (**b**) encrypted version with K1, (**c**) encrypted version with K11

The first group of computer simulation is devoted to evaluate the powerful and efficiency of the proposed technique using the standard Woman image in Matlab. The testing using this image contains many scenarios.

First, test multi secret key and showing the encrypted version of every key. The long of secret key means more complexity of the encryption step, so, May there is no need to complex key or not. It is introduced in this work as optional like embed encrypted medical image or embed the medical image without encryption. The key also, controls the degree of randomizing image pixels. It increases the noise resist, this part is tested also, in the following sections of the simulation results.

As shown in Fig. 7.44, the different of encrypted version of Woman image is produced. The Keys from left to right are numbered as K1, K2, K3,. . ., K16.

With applying the multi-Key on the MRI of Brain medical image, the encrypted versions of the MRI image are given in Fig. 7.45.

7.3.8.2 Evaluation Standard Based

The first section of the computer experiments simulation of the proposed techniques is devoted to evaluate this presented multi-level security technique. The following group of results in this section utilizes the standard Matlab images before applying the technique on the practically medical images.

The standard woman Matlab image is employed in the following experiments. Figure 7.46 gives the original woman image and the encrypted version using K1 and using K11 [$S_{key11} = 64,128,64$].

The last way to view the powerful of the presented technique is cleared in the previous Figs. 7.46, 7.47, 7.48, and 7.49. These figures show the PSNR and the Cr (Correlation coefficient) of the previous scenarios with the SNR channel variation. Figure 7.50 shows the correlation between the extracted images samples and the original standard woman image with the channel SNR variation. This figure also, shows the effects variations of the different scenarios which include the encoded

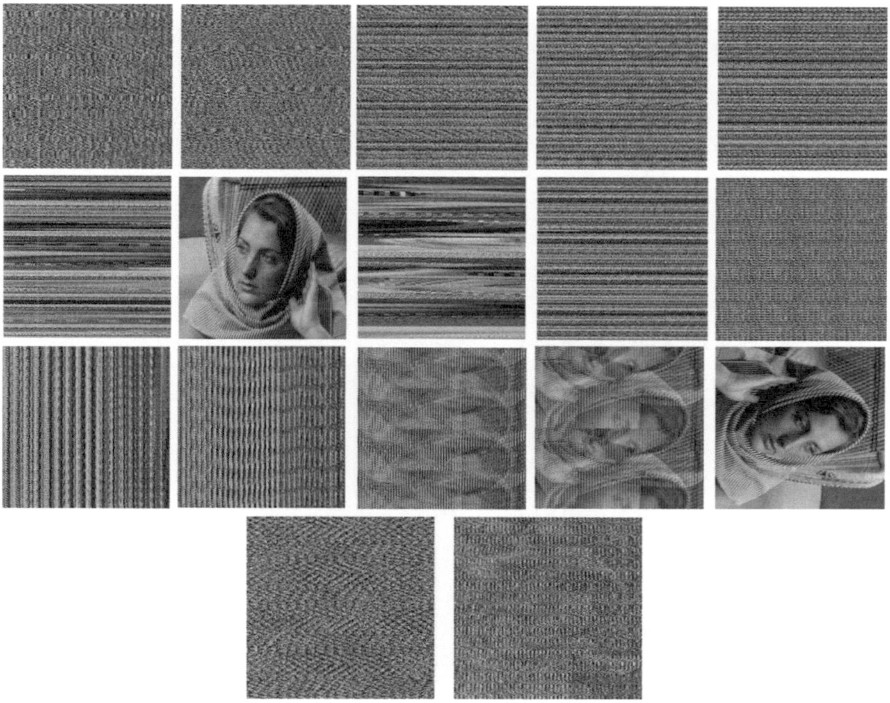

Fig. 7.44 Woman image encrypted versions with different K

and uncoded packets of the watermarked signal transmission. Also, it shows the touch of the encrypted watermark image.

At the lower SNR of channel, the scenarios performance and the extracted images quality from different scenarios are closed. With the moderate and high SNR, the FEC enhances the correlation between the extracted and the original images.

Also, the following Fig. 7.51 gives the PSNR of the extracted standard and encrypted watermark woman images. The PSNR is used to measure the quality of the extracted images metric with the different scenarios with the channel SNR variations.

Figures 7.50 and 7.51 are presented to evaluate and measure the quality of extracted images and show the effectiveness of the proposed technique options, there are encryption process or not and the transmitted packets of the watermarked signal are encoded using the error control schemes or not. Figure 7.52 gives the effectiveness of these options on the whole watermarked signal. It gives the correlation between the transmitted watermarked signal and the received one before the process of extraction of the embedded image.

From the figure, it is cleared that the quality of extracted the embedded image enhances with the encryption utilization also, the degree of improvement increases

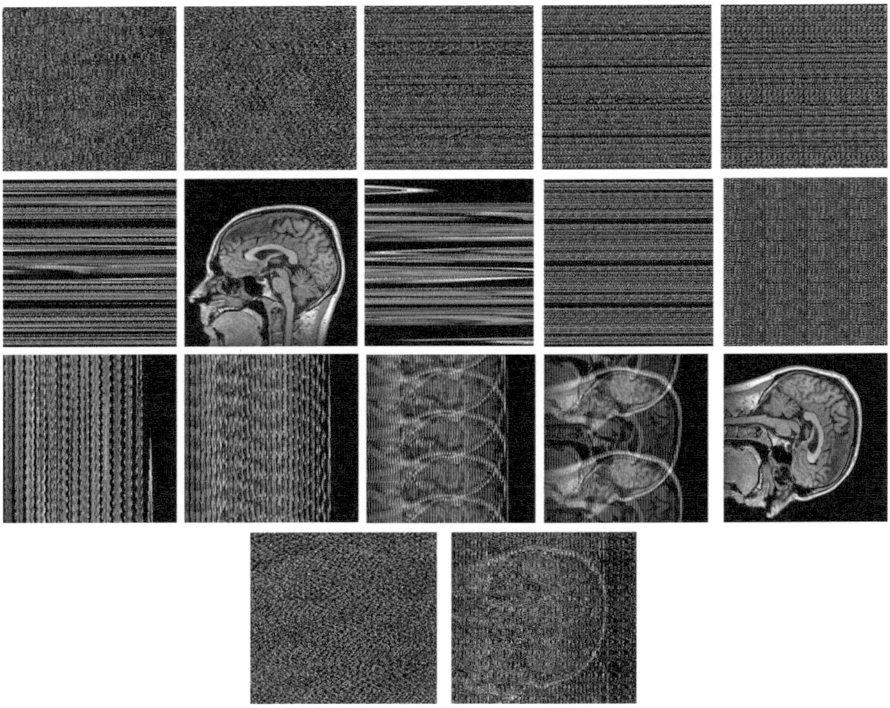

Fig. 7.45 MRI image encrypted versions with different K

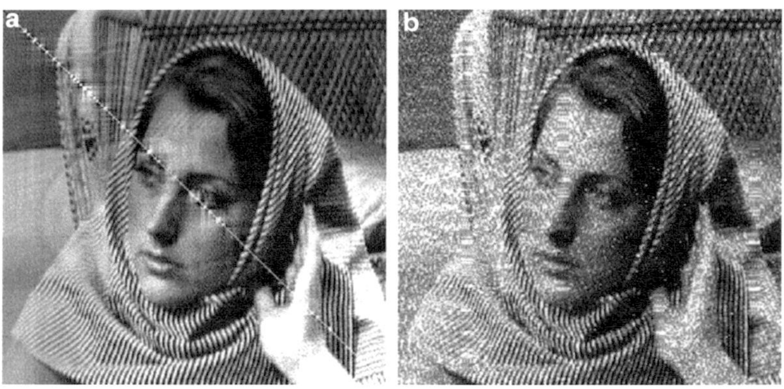

Fig. 7.46 Watermark woman image at SNR = 15 dB using ($2DH_1$) over correlated fading channel. (**a**) The extracted the standard image Cr = 0.072. (**b**) The extracted the encrypted image, Cr = 0.092

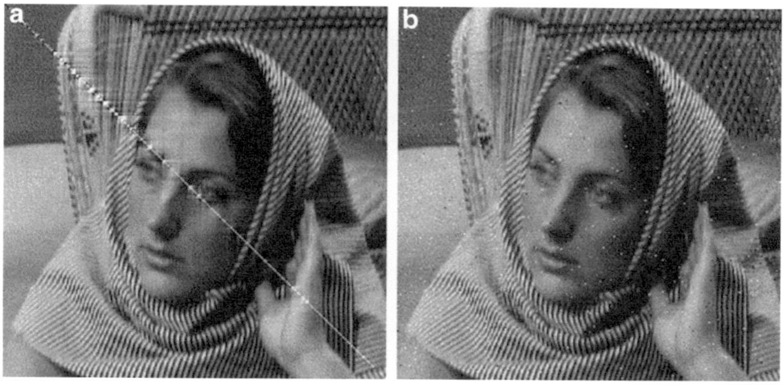

Fig. 7.47 Watermark woman image at SNR = 30 dB using ($2DH_1$) over correlated fading channel. (**a**) The extracted the standard image Cr = 0.64. (**b**) The extracted the encrypted image, Cr = 0.76

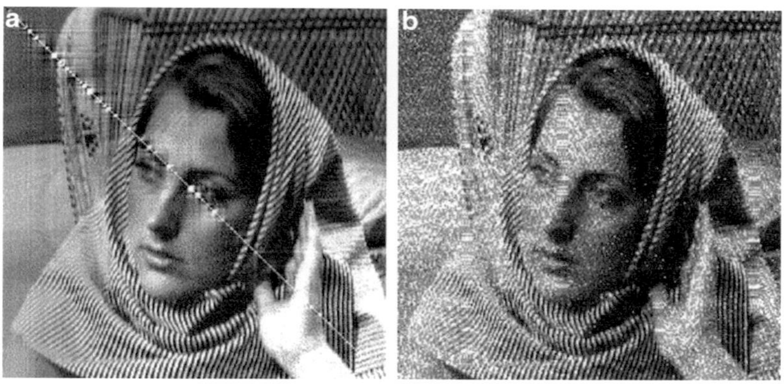

Fig. 7.48 Watermark woman image at SNR = 15 dB using ($2DM_1$) over correlated fading channel. (**a**) The extracted the standard image Cr = 0.079. (**b**) The extracted the encrypted image, Cr = 0.097

with the encoded watermarked signal packets transmission. The effectiveness of the different scenarios is cleared with the higher SNR of channel and the moderate values.

The previous Figs. 7.50 and 7.51 are devoted to evaluate the proposed technique with its different scenarios through measuring the extracted image quality using standard Matlab woman image. In the next figure, the watermarked signal behavior is studied and its performance changing with the previous different scenarios. The watermarked signal is the combined transmitted signal; it is the audio signal with the embedded image. As shown in Fig. 7.52, the encryption utilization for the embedded image is not affect on the watermarked signal, it is expected and logic results. The FEC scheme introduces little improving more than the uncoded packets transmission [36].

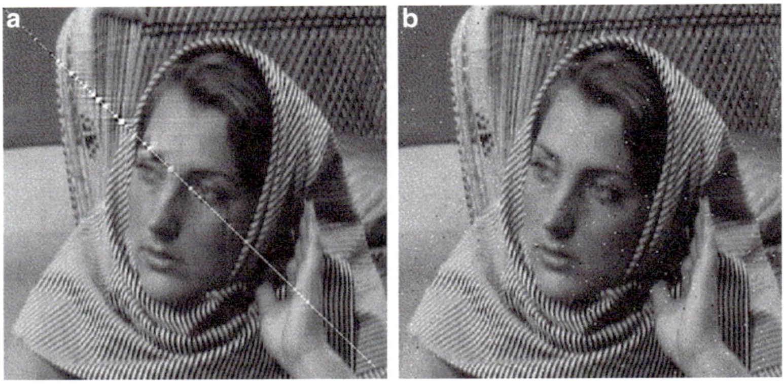

Fig. 7.49 Watermark woman image at SNR = 30 dB using ($2DM_1$) over correlated fading channel. (**a**) The extracted the standard image Cr = 0.86. (**b**) The extracted the encrypted image, Cr = 0.92

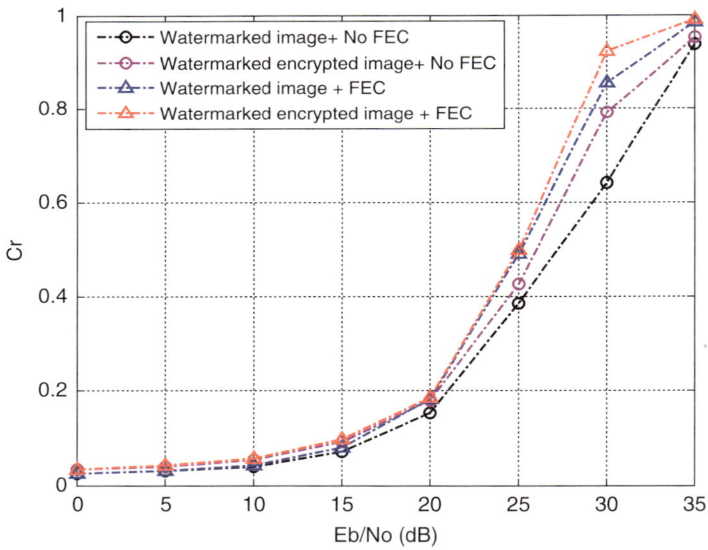

Fig. 7.50 Cr between the transmitted image and the extracted images also, the embedded audio watermark with the received ones

7.3.8.3 Evaluation of X-Ray Images

In this section of experiments, the first medical image is utilized to evaluate the presented technique on the medical image. The simulations computer experiments are carried out over the AWGN channel with different scenarios.

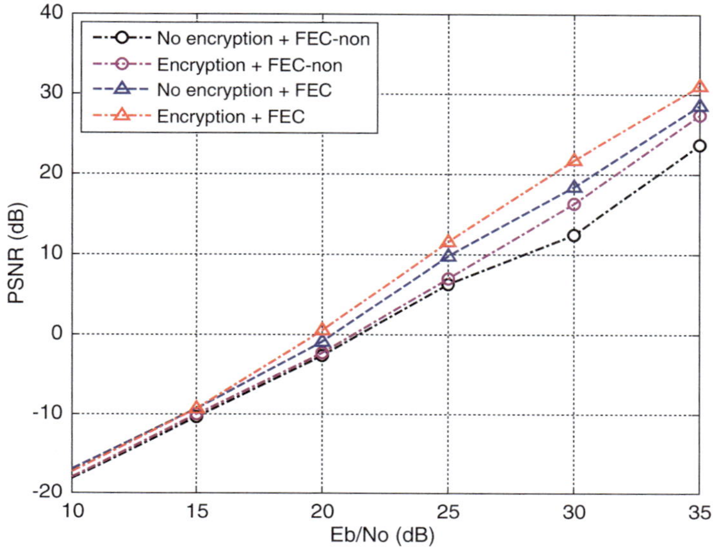

Fig. 7.51 The PSNR with E_b/No variation with different scenarios

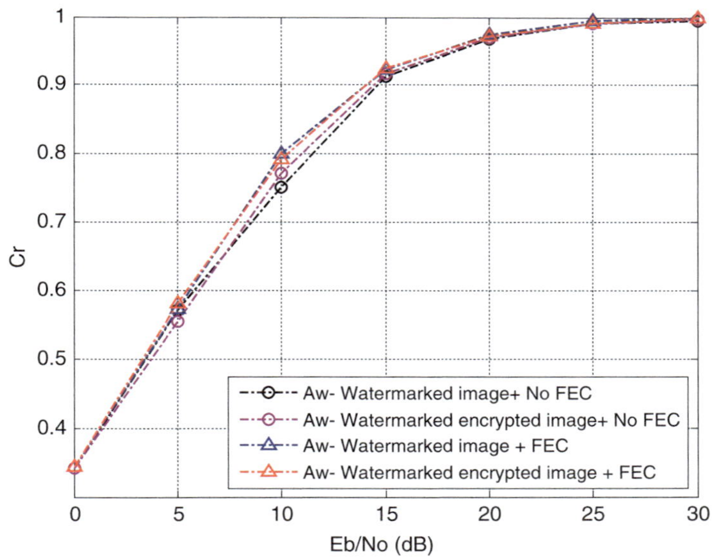

Fig. 7.52 Cr between the transmitted audio watermarked signal and the received one with the different scenarios, encryption and FEC

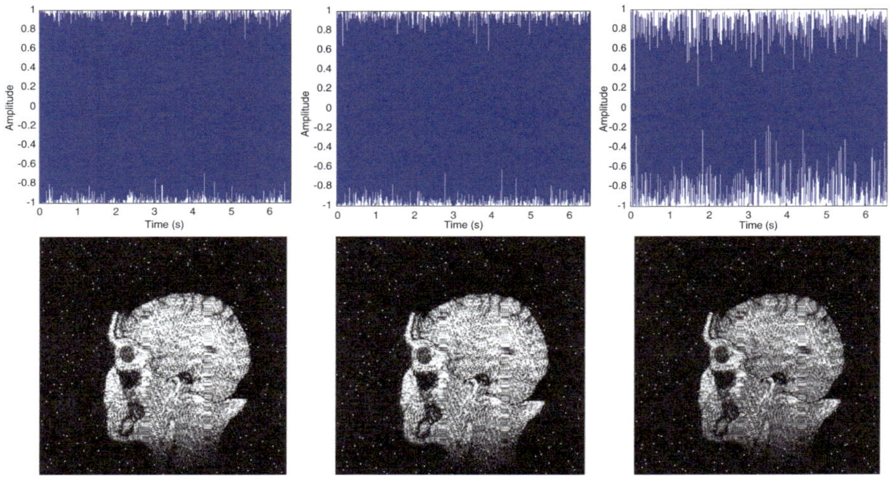

Fig. 7.53 Gives the results of X-Ray image samples SNR = 0 dB AWGN with different FEC, (**a**) No FEC, (**b**) simple FEC, (**c**) powerful FEC

Figure 7.53 gives the results of X-Ray inspection image of the human head samples at the channel SNR = 0 dB over the AWGN channel. Figure 7.53 shows the received audio watermarked signal in the first row and the second one show the extracted of the encrypted images. The columns show the received audio watermarked signal and the extracted medical images in case of different scenarios, as follows from the left to right, scenario I = the encrypted embedded medical X-ray image without FEC utilization, scenario II = the encrypted image with the simple and weak error control scheme Hamming (15,11) code and the last scenario III = the encrypted image with another powerful FEC scheme the convolutional code (1, 2, 3).

Figure 7.54 shows the results of watermarking signal and the extracted X-ray image over the AWGN channel with the channel SNR = 5 dB. As shown in this Fig. 7.54, there are no good effects of weak FEC especially at lower SNR of channel.

7.3.8.4 Evaluation MRI Images

In this section of experiments, the inspection medical MRI image is utilized to evaluate the presented technique on the medical image. The simulations computer experiments are carried out over the uncorrelated fading channel with the previous different scenarios.

Figure 7.55 shows the results extracted MRI images samples using the uncorrelated fading channel with the channel SNR variation.

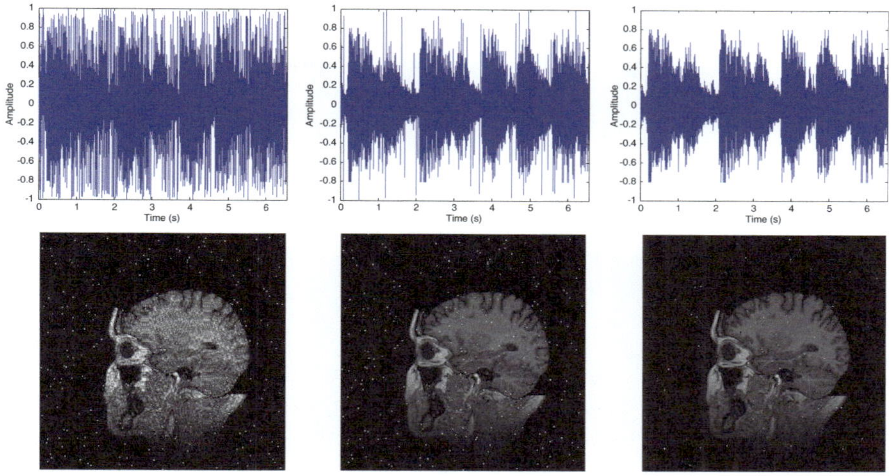

Fig. 7.54 Gives the results of X-Ray image samples SNR = 5 dB AWGN with different FEC, (**a**) No FEC, (**b**) simple FEC, (**c**) powerful FEC

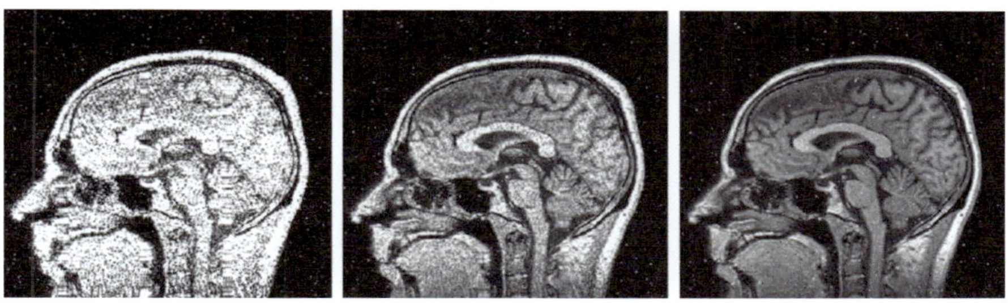

Fig. 7.55 Extracted watermarks after transmission with Bluetooth network using $2DH_1$ uncoded packets over fading channel. (**a**) SNR = 5 dB, (**b**) SNR = 15 dB and, (**c**) SNR = 25 dB

Figure 7.56 gives the received audio watermarking signal, the extracted encrypted MRI image and the extracted MRI image after decryption with different SNR using weak FEC.

To enhance the extracted MRI image and evaluate it with using powerful FEC to encode the transmitted packets the Convolutional code with R = 1/2 and k = 1 is utilized within this experiments. Figure 7.57 gives the received audio watermarking signal, the extracted encrypted MRI image and the extracted MRI image after decryption with different SNR using powerful FEC.

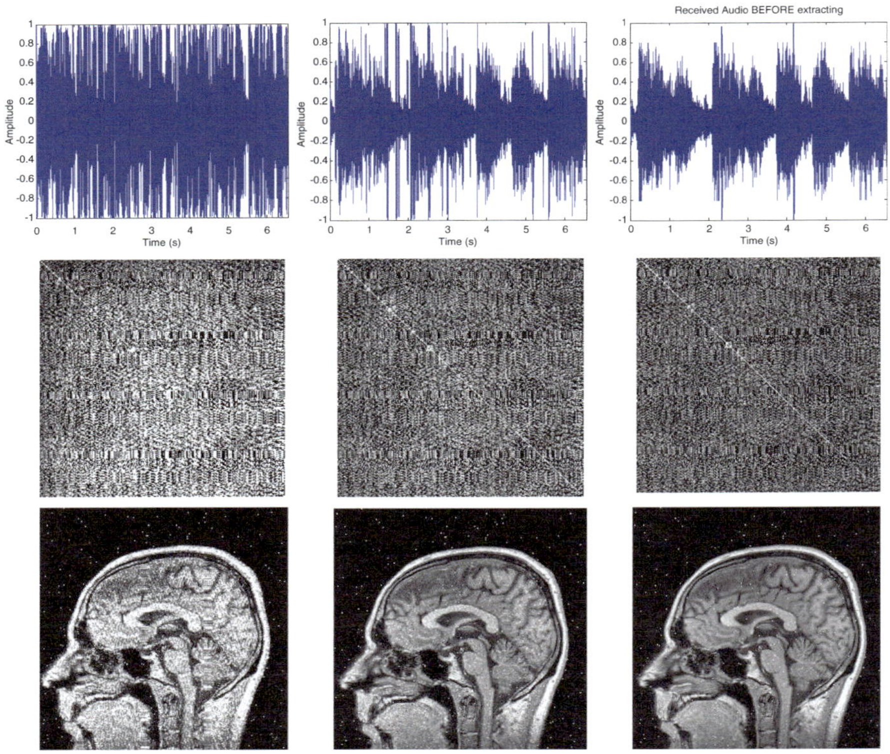

Fig. 7.56 Extracted watermarks after transmission with Bluetooth network using $2DM_1$ encoded packets with Hamming (15, 10) code over fading channel. (**a**) SNR = 5 dB, (**b**) SNR = 15 dB, and (**c**) SNR = 25 dB

7.3.8.5 Experiment of $2DH_3$ and $2DM_3$ Packets

In this section of the experiment, the variation of packet length transmission of the watermarked signal is studied within many scenarios. This experiment group utilizes different packet length to show extracted medical images quality. In the following the results of the experiments are tabulated in Table 7.2 [29].

Table 7.2 tabulates the results of these section experiments; it shows the relation between the length of transmitted packets and its formats and the quality of the extracted MRI medical images. Also, Fig. 7.58 shows the samples of the received audio watermarking signals and the extracted MRI image with the different packets format over the correlated fading channel.

In the following, Fig. 7.59 gives the experiment of the embedding the encrypted MRI image within Audio signals with the FEC variation and longer packet segmentation is presented. This experiment is devoted to investigate the longer packets

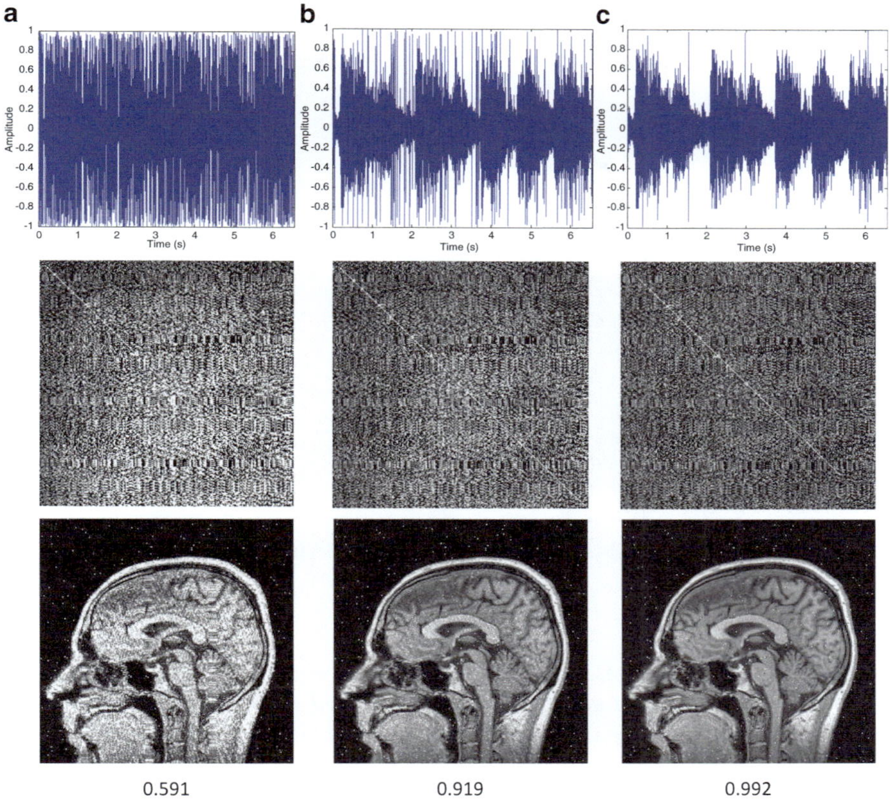

Fig. 7.57 Extracted watermarks after transmission with Bluetooth network using $2DM_1$ encoded packets with Convolutional (R = 1/2 and K = 3) code over fading channel. (**a**) SNR = 5 dB, (**b**) SNR = 15 dB, and (**c**) SNR = 25 dB

Table 7.2 Results of different length of packets Cr of the extracted image

Jakes model with the Vc = 20 m/h and SNR = 15 dB secret key = K_1					
Pkt	Cr	Pkt	Cr	Pkt	Cr
2DH1	0.911	2DM1	0.923	2DM1_Conv123	0.964
2DH3	0.9095	2DM3	0.9188	2DM3_Conv123	0.963
2DH3_K_{11}	0.9034	2DM3_K_{11}	0.9166	2DM3_Conv123_K_{11}	0.960

transmission effects on the extracted medical MRI-image of human brain. The FEC schemes are the same FEC in the previous experiment. The channel SNR is 15 dB.

As shown in the proposed scenarios section, the powerful of the presented technique applying within the medical image processing is based on its flexibility of its secret key. This key is not secret key may be changed to provide additional security level for the transmitted information, only it may be employed to decrease

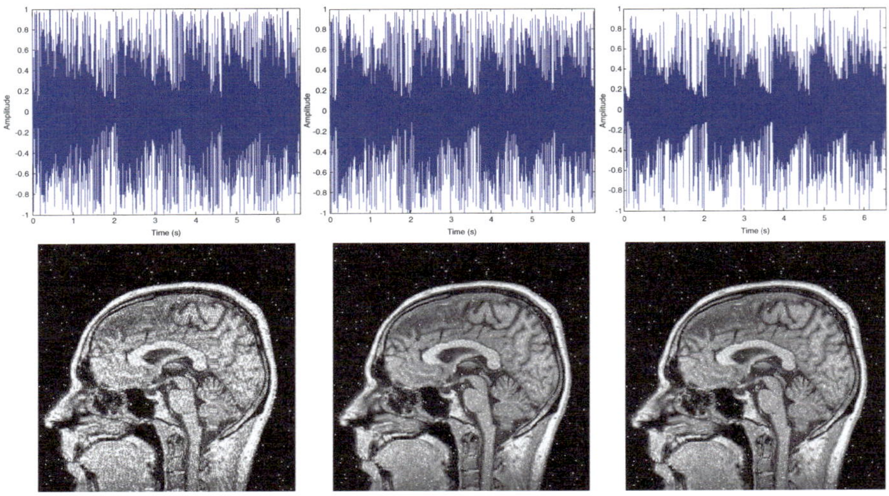

Fig. 7.58 The extracted MRI image with the scenarios of FEC schemes (packet length = 512 bits). (**a**) Encrypted MRI image with No FEC, (**b**) simple FEC, (**c**) powerful FEC

noise channel. It lets the transmitted data resist the channel effects if it is chosen correctly. The encryption algorithm of the embedded medical images complexity depends on the secret key also.

The next experiment is carried out to test the role of secret key in this issue. The K11 ($S_{key11} = [64,128,64]$ is utilized in this experiment as just example. Figure 7.60 gives the results of the MRI image samples.

7.3.9 Scenario Recommendation

This scenario presented an efficient SVD audio watermarking approach. In this approach, encrypted images are embedded as watermarks in audio signals to achieve a higher degree of security than traditional audio watermarking algorithms. The proposed audio watermarking approach can be implemented on the audio signal as a whole or on a segment-by-segment basis. Experimental results have proved that watermark embedding in the proposed approach does not deteriorate the audio signals. It has been clear through experiments that the chaotic Baker map encryption algorithm is an efficient algorithm for watermark encryption.

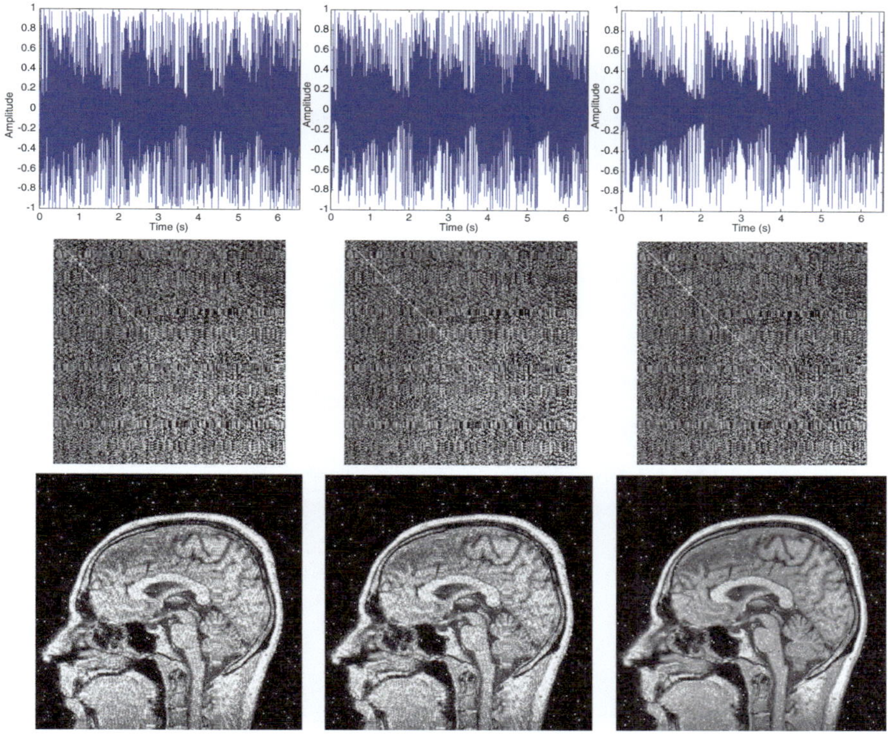

Fig. 7.59 The extracted MRI image with the scenarios of FEC schemes (packet length = 1,500 bits). (**a**) Encrypted MRI image with No FEC, (**b**) simple FEC, (**c**) powerful FEC

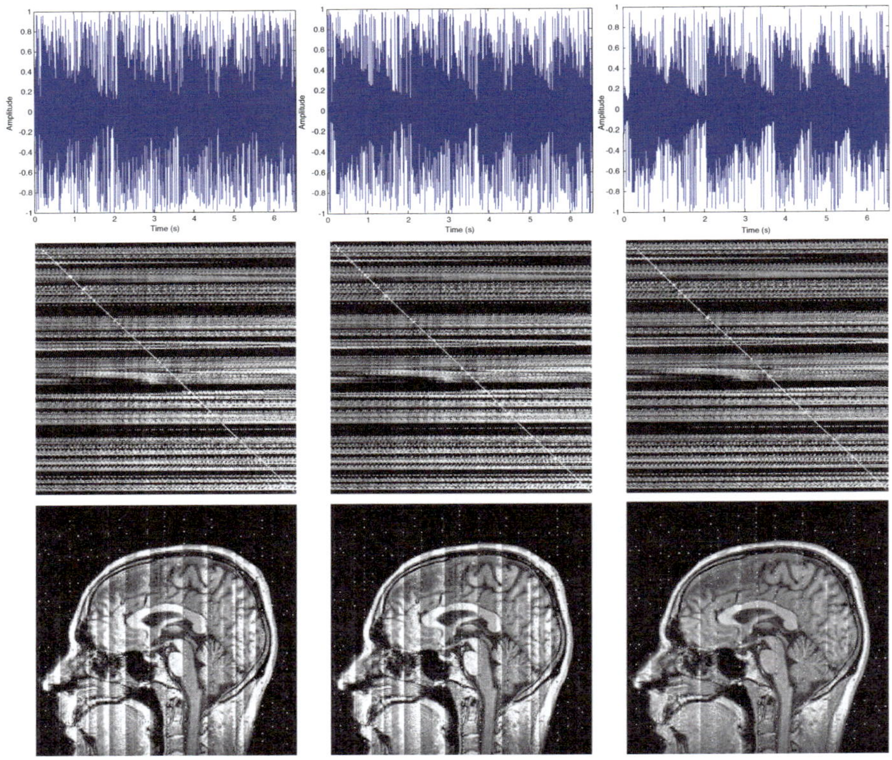

Fig. 7.60 The extracted of encrypted MRI Image with K11 the scenarios of FEC schemes (packet length = 1,500 bits). (**a**) Encrypted MRI image with No FEC, (**b**) simple FEC, (**c**) powerful FEC

References

1. Sklavos N, Zhang X (2007) Wireless security & cryptography: specifications and implementations. CRC Press/Taylor and Francis. ISBN:084938771X
2. Macq B, Dittmann J, Delp EJ (2004) Benchmarking of image watermarking algorithms for digital rights management. Proc IEEE 92(6):971–984
3. Lu ZM, Xu DG, Sun SH (2006) Multipurpose image watermarking algorithm based on multistage vector quantization. IEEE Trans Image Process 14(6):822–831
4. Kim HS, Lee HK (2003) Invariant image watermark using Zernike moments. IEEE Trans Circuit Syst Video Technol 13(8):766–775
5. Chu WC (2003) DCT-based image watermarking using subsampling. IEEE Trans WSN 5 (1):34–38
6. Ghouti L, Bouridane A, Ibrahim MK, Boussakta S (2006) Digital image watermarking using balanced multiwavelets. IEEE Trans Signal Process 54(4):1519–1536
7. Xiang S, Huang J (2007) Histogram-based audio watermarking against time-scale modification and cropping attacks. IEEE Trans WSN 9(7):1357–1372

8. Liu Z, Inoue A (2003) Audio watermarking techniques using sinusoidal patterns based on pseudorandom sequences. IEEE Trans Circuit Syst Video Technol 13(8):801–812
9. Lemma AN, Aprea J, Oomen W, de Kerkhof LV (2003) A temporal domain audio watermarking technique. IEEE Trans Signal Process 51(4):1088–1097
10. Li W, Xue X, Lu P (2006) Localized audio watermarking technique robust against time-scale modification. IEEE Trans WSN 8(1):60–69
11. Erküçük S, Krishnan S, Glu MZ (2006) A robust audio watermark representation based on linear chirps. IEEE Trans WSN 8(5):925–936
12. Wang X, Qi W, Niu P (2007) A new adaptive digital audio watermarking based on support vector regression. IEEE Trans Audio Speech Lang Process 15(8):2270–2277
13. Özer H, Sankur B (2005) An SVD based audio watermarking technique. In: Proceedings of the IEEE 13th conference on signal processing and communications applications, pp 452–455
14. Han F, Yu X, Han S (2006) Improved Baker map for image encryption. In: Proceedings of the first international symposium on systems and control in aerospace and astronautics, ISSCAA 2006, pp 1273–1276
15. Fridrich J (1997) Image encryption based on chaotic maps. In: Proceedings of the IEEE international conference on systems, man, and cybernetics, pp 1105–1110
16. Qian Q, Chen Z, Yuan Z (2008) Video compression and encryption based-on multiple chaotic system. In: Proceedings of The 3rd international conference on innovative computing information and control (ICICIC'08)
17. Huang F, Lei F (2008) A novel symmetric image encryption approach based on a new invertible two-dimensional map. In: Proceedings of the international conference on intelligent information hiding and WSN signal processing IIHMSP 2008, pp 1340–1343
18. Koduru SC, Chandrasekaran V (2008) Integrated confusion-diffusion mechanisms for chaos based image encryption. In: Proceedings of the IEEE 8th international conference on computer and information technology workshops, pp 260–263
19. Usman K, Juzojil H, Nakajimal I (2007) Medical image encryption based on pixel arrangement and random permutation for transmission security. In: Proceedings of the 9th international conference on e-health networking, application and services, Taipei, pp 244–247
20. Elashry IF, Farag Allah OS, Abbas AM, El-Rabaie S, Abd El-Samie FE (2009) Homomorphic image encryption. J Electron Imaging 18(3):033002, SPIE
21. Ghazy RA, Hadhoud MM, Dessouky MI, El-Fishawy NA, Abd El-Samie FE (2008) Performance evaluation of block based SVD image watermarking. Progr Electromagn Res B 8:147–159
22. Liu R, Tan T (2002) An SVD-based watermarking scheme for protecting rightful ownership. IEEE Trans WSN 4(1):121–128
23. Sun X, Liu J, Sun J, Zhang Q, Ji W (2008) A robust image watermarking scheme based-on the relationship of SVD. In: Proceedings of the international conference on intelligent information hiding and WSN signal processing, pp 731–734
24. Zhu X, Zhao J, Xu H (2006) A digital watermarking algorithm and implementation based on improved SVD. In: Proceedings of the IEEE 18th international conference on pattern recognition (ICPR'06), pp 651–656
25. Kubichek R (1993) Mel-cepstral distance measure for objective speech quality assessment. In: Proceedings of the IEEE Pacific Rim conference on communications, computers and signal processing, pp 125–128
26. Cordeiro CM, Abhyankar S, Toshiwal R, Agrawal DP (2004) BlueStar: enabling efficient integration between Bluetooth WPANs and IEEE 802.11 WLANs. Mobile Netw Appl 9:409–422
27. Chen J-J, Sun T, Chen Y-C (2006) Improving Bluetooth throughput using FEC and interleaving. In: International conference on mobile ad hoc sensor networks, Hong Kong, pp 726–736
28. Waleed GM, Faizal M, Ahmed RB, Fareq M, Abd Malek B, Alif Ghani M (2009) Bluetooth wireless network authentication using radio frequency communication protocol. J Comput Sci 5(9):649–650

29. Kumar T (2009) Improving pairing mechanism in Bluetooth security. Int J Recent Trends Eng 2(2)
30. Albert J, Bissyandé TF, Bromberg YD, Chaumette S, Réveillère L (2010) UbiPAN: a Bluetooth extended personal area network. In: 2010 international conference on complex, intelligent and software intensive systems, pp 774–778
31. Wang S, Sekey A, Gersho A (2002) An objective measure for predicting subjective quality of speech coders. IEEE J Sel Areas Commun 10(5):819–829
32. Yang W, Benbouchta M, Yantorno R (1998) Performance of the modified bark spectral distortion as an objective speech quality measure. In: Proceedings of the IEEE international conference on acoustic, speech and signal processing (ICASSP), Washington, vol 1, pp 541–544
33. McDermott BJ, Scaglia C, Goodman DJ (1978) Perceptual and objective evaluation of speech processed by adaptive differential PCM. In: Proceedings of the IEEE international conf. on acoustic, speech and signal processing (ICASSP), Tulsa, pp 581–585
34. Crochiere RE, Tribolet JE, Rabiner LR (1980) An interpretation of the log likelihood ratio as a measure of waveform coder performance. IEEE Trans Acoust Speech Signal Process Assp-28(3):318–323
35. Kai B, Yong P (2006) Performance study on Bluetooth-based wireless personal area networks for real-time health monitoring. ETRI J 28(4)
36. Lai C-C (2011) An improved SVD-based watermarking scheme using human visual characteristics. Optics Commun Sci Direct 284:938–944

MIX
Papier aus verantwortungsvollen Quellen
Paper from responsible sources
FSC® C105338
FSC
www.fsc.org

If you have any concerns about our products,
you can contact us on
ProductSafety@springernature.com

In case Publisher is established outside the EU,
the EU authorized representative is:
Springer Nature Customer Service Center GmbH
Europaplatz 3, 69115 Heidelberg, Germany

Printed by Libri Plureos GmbH
in Hamburg, Germany